POWER GRID OPERATION IN A MARKET ENVIRONMENT

POWER GRID OPERATION IN A MARKET ENVIRONMENT

Economic Efficiency and Risk Mitigation

Edited by

HONG CHEN

IEEE
PRESS
SERIES
ON POWER
ENGINEERING

IEEE PRESS

WILEY

Published by John Wiley & Sons, Inc., Hoboken, New Jersey.
Published simultaneously in Canada.

For general information on our other products and services or for technical support, please contact our Customer Care Department within the United States at (800) 762-2974, outside the United States at (317) 572-3993 or fax (317) 572-4002.

Wiley also publishes its books in a variety of electronic formats. Some content that appears in print may not be available in electronic formats. For more information about Wiley products, visit our web site at www.wiley.com.

Library of Congress Cataloging-in-Publication Data is available.

ISBN: 978-1-118-98454-3

Printed in the United States of America.

10 9 8 7 6 5 4 3 2 1

CONTENTS

FOREWORD   ix

PREFACE   xi

ACKNOWLEDGMENT   xiii

CONTRIBUTORS   xv

PART I   *INTEGRATED SYSTEM AND MARKET OPERATION*

CHAPTER 1   *BALANCE ECONOMIC EFFICIENCY AND OPERATION*
             RISK MITIGATION   3
             Hong Chen and Jianwei Liu

1.1   Power System Operation Risk Mitigation: The Physics   **4**
1.2   Integrated System and Market Operation: The Basics   **11**
1.3   Economic Efficiency Evaluation and Improvement: The Economics   **20**
1.4   Final Remarks   **35**
      Appendix 1.A Nomenclature   **36**
      Appendix 1.B Electricity Market Model   **37**
      References   **39**
      Disclaimer   **41**

CHAPTER 2   *MITIGATE MARKET POWER TO IMPROVE MARKET EFFICIENCY*   43
             Ross Baldick

2.1   Introduction   **43**
2.2   Price Formation in Electricity Markets   **50**
2.3   Price and Offer Caps   **52**
2.4   Ability and Incentive to Exercise Market Power   **53**
2.5   Market Power Mitigation Approaches   **57**
2.6   Conclusion   **65**
      Acknowledgments   **65**
      References   **65**

PART II   *UNDER SMART GRID ERA*

CHAPTER 3   *MASS MARKET DEMAND RESPONSE MANAGEMENT FOR*
             THE SMART GRID   69
             Alex D. Papalexopoulos

3.1 Overview **69**
3.2 Introduction **72**
3.3 Distributed Computing-Based Demand Response Management Approach **74**
3.4 The ColorPower Architecture and Control Algorithms **75**
3.5 Integration with the Wholesale Energy Market **80**
3.6 Equalizing Market Power Between Supply and Demand **83**
3.7 Generalization Beyond Demand Response **84**
3.8 A Numerical Example **87**
3.9 Concluding Remarks **88**
   Appendix 3.A Nomenclature **89**
   References **89**

CHAPTER 4 *IMPROVE SYSTEM PERFORMANCE WITH LARGE-SCALE*
VARIABLE GENERATION ADDITION **91**
Yuri V. Makarov, Pavel V. Etingov, and Pengwei Du

4.1 Review of Regulation and Ancillary Services **92**
4.2 Day-Ahead Regulation Forecast at CAISO **93**
4.3 Ramping and Uncertainties Evaluation at CAISO **99**
4.4 Quantifying the Regulation Service Requirements at ERCOT **103**
4.5 Conclusions **111**
   Appendix 4.A Nomenclature **112**
   References **113**

PART III *STOCHASTIC APPLICATIONS*

CHAPTER 5 *SECURITY-CONSTRAINED UNIT COMMITMENT WITH*
UNCERTAINTIES **117**
Lei Wu and Mohammad Shahidehpour

5.1 Introduction **118**
5.2 SCUC **119**
5.3 Uncertainties in Emerging Power Systems **125**
5.4 Managing the Resource Uncertainty in SCUC **134**
5.5 Illustrative Results **155**
5.6 Conclusions **163**
   Appendix 5.A Nomenclature **164**
   Acknowledgments **166**
   References **166**

CHAPTER 6 *DAY-AHEAD SCHEDULING: RESERVE DETERMINATION*
AND VALUATION **169**
Ruiwei Jiang, Antonio J. Conejo, and Jianhui Wang

6.1 The Need of Reserves for Power System Operation **169**
6.2 Reserve Determination via Stochastic Programming **170**
6.3 Reserve Determination via Adaptive Robust Optimization **179**
6.4 Stochastic Programming vs. Adaptive Robust Optimization **182**
6.5 Reserve Valuation **185**

6.6   Summary, Concluding Remarks, and Research Needs   **191**
       Appendix 6.A Nomenclature   **192**
       References   **193**

PART IV *HARNESS TRANSMISSION FLEXIBILITY*

CHAPTER 7   *IMPROVED MARKET EFFICIENCY VIA TRANSMISSION*
               SWITCHING AND OUTAGE EVALUATION IN SYSTEM OPERATIONS   **197**
               Kwok W. Cheung and Jun Wu

7.1   Background   **197**
7.2   Basic Dispatch Model for Market Clearing   **198**
7.3   Economic Evaluation of Transmission Outage   **201**
7.4   Optimal Transmission Switching   **203**
7.5   Selection of Candidate Transmission Lines for Switching and Implementation
       of OTS   **206**
7.6   Test Cases   **210**
7.7   Final Remarks   **216**
       Appendix 7.A Nomenclature   **216**
       References   **217**

CHAPTER 8   *TOWARD VALUING FLEXIBILITY IN TRANSMISSION PLANNING*   **219**
               Chin Yen Tee and Marija D. Ilić

8.1   Introduction   **219**
8.2   Scale Economies of Transmission Technologies   **221**
8.3   Disconnect of Current Power System Operational, Planning, and
       Market Mechanisms   **225**
8.4   Impact of Operational and Market Practices on Investment Planning   **225**
8.5   Information and Risk Sharing in the Face of Uncertainties   **230**
8.6   Challenges in Designing Financial Rights for Flexibility   **234**
8.7   Conclusions   **235**
       Appendix 8.A Nomenclature   **236**
       Appendix 8.B Mathematical Models Used for Case Studies   **238**
       Appendix 8.C Investment Cost   **247**
       References   **248**

INDEX   **251**

FOREWORD

SINCE THE INTRODUCTION of deregulated energy markets, the debate on reliability versus efficiency has continued. As many of the chapter contributors in this edited book, *Power Grid Operation in a Market Environment: Economic Efficiency and Risk Mitigation*, referenced, grid reliability continues to be a priority throughout established markets and regulated grid operations. While the reliability of power systems clearly remains the highest priority, the need for these operations to achieve an overall efficient use of resources must also remain a priority. This book's contributing authors tackle that aspect by addressing the existing efficiency challenges and introduce innovative approaches to continue to improve efficiency while maintaining reliability.

The book is a comprehensive technical handling of many of the challenges that deregulated grid operations face and must continue to improve upon. The book does not attempt to provide all of the technical possibilities, but instead identifies some key issues and offers some innovative approaches to the primary challenges that face grid operators.

The book contributors have worked for several years from a collaboration of one of the IEEE Power Energy Society teams focused on this topic. The book will be valuable as both an academic reference as well as a reference work for market system design. Each of the chapters takes an in-depth look at a particular aspect of efficient market design.

Chen and Liu present an excellent overview of system operation and market design in Chapter 1 including the stages of mature market design to include multi-settlement markets, capacity markets, and risk-based and dynamic markets. These authors also address the first efficiency aspect—efficiency effects of reliable grid operations.

In Chapter 2, Baldick tackles another aspect of efficiency in market power and market mitigation—in general, those inefficiencies caused by imperfect markets and competition. This chapter discusses some of the short-term and long-term inefficiencies and provides a nice tie-in to the effects of demand response (DR) on market power, serving as an introduction to the next chapter. In Chapter 3, Papalexopoulos discusses the impact of demand response resources on current market operations, and looks at a more efficient approach to handling the deployment of demand response resources. This chapter describes some of the current inefficiencies in DR deployment and the necessity of a new distributed approach to address the increase of these resources in grid operations.

A natural follow-on to the demand response discussion is the proliferation of intermittent resources and their impact to efficient market and grid operations.

Makarov, Etingov, and Du tackle the increase of renewable resources and their effect on system performance. These authors offer several solutions for grid operators to incorporate the uncertainty of these renewable resources into the efficient dispatch for grid operations. This sets up a good introduction for the next chapter where Wu and Shahidehpour discuss unit commitment challenges in Chapter 5. Unit commitment continues to be a challenging algorithmic problem in organized markets, particularly when you look at the impact of demand response, renewables, and the optimization of energy and reserve commitment. This chapter presents several potential technical approaches to solve these challenges.

In Chapter 6, Jiang, Conejo, and Wang take the handoff and focus on the particular challenge of reserve requirements and how reserve commitments are valued in market operations. The approach in this chapter is unique as it is presented as a tutorial on reserve commitment with several approaches to adding additional efficiency to the determination and deployment of reserves.

The final two chapters take a different direction while completing the circle in addressing the various aspects of efficient operations. In chapter 7, Cheung and Wu tackle market efficiency affected by transmission outages and transmission-switching opportunities. The authors discuss the current approach which assumes a fixed network topology, and explore several methods to improve efficient operations through more dynamic approaches to both transmission outage scheduling and transmission switching. These approaches include the recognition of potential affects to forward market operations. The chapter provides a good transition to the final chapter where Tee and Ilić present a planning framework to account for the value of operational flexibility in transmission planning.

Overall, the book presents a holistic look at the various operations, markets, and transmission planning challenges to improving efficient operations. The potential solutions and approaches detailed in this book will be a good foundation to future improvements and debates on improving grid and market efficiency.

Michael Bryson
Vice President of Operations
PJM Interconnection
Philadelphia, USA

PREFACE

ELECTRICITY MARKETS HAVE BEEN successfully implemented in many parts of the world, and are reshaping power grid operations philosophically and practically. Economic efficiency has become one of the important objectives of grid operation and planning, along with the fundamental responsibilities of operation risk mitigation. Market and system operators are facing challenges to define and achieve the equilibrium of economic efficiency and risk mitigation.

This edited book, *Power Grid Operation in a Market Environment: Economic Efficiency and Risk Mitigation*, covers both system operation and market operation perspective, especially focusing on the interaction between the two. It reveals the challenges and best practices of the industry, and also presents the latest researches on this topic, which helps us to better prepare for the challenges and new trends in the industry.

The overview of integrated system and market operation is provided in Chapter 1, discussing the integrated operation philosophy, current market design, practices and challenges, as well as PJM's (a RTO/ISO in United States, in charge of 13 states and DC area's bulk power system operation and planning, and operating the largest wholesale electricity market in the world) experience on evaluating and improving economic efficiency. Often and conveniently, economic efficiency is based on bid and offer prices, assuming competitive markets. However, the practical electricity markets are not fully competitive. Systematic methods are needed to mitigate market power to improve market efficiency, which is discussed in Chapter 2.

With demand participation in the market mechanism and smart grid technologies, system demand becomes more elastic, which significantly improves economic efficiency and also helps with operation risk mitigation. Chapter 3 describes a new approach for the mass market demand response management, which is an essential component of the smart grid infrastructure. On the supply side, more and more renewable resources are being integrated to the system, which inevitably introduces high system volatility, sometimes, can cause high frequency excursion, as well as reduced inertia response impacting transient stability, especially for small systems. Therefore, more flexible resources are needed for system operation to improve system performance. Chapter 4 introduces new criteria to improve system performance with large-scale variable generation addition.

Operational uncertainties are challenging to manage, and significantly impact economic efficiency, especially with the increased uncertainties brought by high penetration of renewable resources. Stochastic applications can help to address the challenges. Mathematic models and solution methods of Security-Constrained Unit Commitment (SCUC) are discussed in detail in Chapter 5, considering various system uncertainties. Reserves are essential to system operation to hedge operation risks caused by uncertainties. Current deterministic methods use fixed reserve requirements determined offline based on historical data and/or procedure, and may not reflect the changing system reliability needs. Most times deterministic methods have higher level of conservativeness than required by the actual conditions, which therefore impacts economic efficiency. Chapter 6 presents a stochastic method to determine and value reserves.

On the network topology side, today's market and system operation mainly assumes fixed network topology. Emerging technologies can improve and utilize flexible transmission control to improve economic efficiency. Chapter 7 describes the methods to improve economic efficiency through topology control, that is, transmission switching and outage scheduling. In parallel, Chapter 8 presents a planning framework to account for the value of operational flexibility in transmission planning and to provide market mechanism for the risk sharing.

In summary, balancing economic efficiency and operation risk mitigation has been an ongoing challenge for the power grid operations in a market environment. It is being addressed from all aspects: from overall market design and system performance to solution methodologies; from supply and demand to networks. All of these contribute toward finding an equilibrium of economic efficiency and risk mitigation for power systems.

The book is a result of more than 5 years' efforts on IEEE Power Energy Society (PES) Task Force "Equilibrium of Electricity Market Efficiency and Power System Operation Risk." It features the most current insight of integrated operation and state-of-the-art development, with field experience and evidence of considerable market savings by tracking equilibrium in operation. It will provide invaluable and timely reference for power engineers, electricity market traders and analysts, market designers and researchers, as well as graduate students, to understand the integrated electricity market and power system operation, reveal new requirements of vendor products, and stimulate new research and development initiatives, especially on modeling and computational techniques in system operation and market analysis. Its unique angle of views to the electricity market and power system operation will be a good compensation to the current literature.

ACKNOWLEDGMENT

I **GIVE MY HEARTFELT THANKS** to all the contributors for their enthusiasm and timely inputs. Without them, this book would never become a reality. I also appreciate all the reviewers for their constructive comments and suggestions, and all the Task Force members for their active participation and discussion. Besides, I'm thankful for the help and patience from the staff of the IEEE Press and Wiley, specially, Dr. Mohamed E. El-Hawary, Editors Mary Hatcher and Brady A. Chin. I'm most grateful for the support my employer, PJM, has been giving me all these years. Last but not least, I am indebted to my family for their encouragement during the book development. My daughters, Sophia and Alice, truly understand my passion and commitment to power and energy society.

Hong Chen
Philadelphia, USA

CONTRIBUTORS

Ross Baldick (F'07) received B.Sc. degree in mathematics and physics and B.E. degree in electrical engineering from the University of Sydney, Sydney, New South Wales, Australia, and M.S. and Ph.D. degrees in electrical engineering and computer sciences from the University of California, Berkeley, CA, USA, in 1988 and 1990, respectively. From 1991 to 1992, he was a Postdoctoral Fellow with the Lawrence Berkeley Laboratory, Berkeley, CA, USA. In 1992 and 1993, he was an Assistant Professor with Worcester Polytechnic Institute, Worcester, MA, USA. He is currently a Professor with the Department of Electrical and Computer Engineering, the University of Texas at Austin, Austin, TX, USA. His research involves optimization, economic theory, and statistical analysis applied to electric power systems, particularly in the context of increased renewables and transmission. Dr. Baldick is a Fellow of the IEEE and the recipient of the 2015 IEEE PES Outstanding Power Engineering Educator Award.

Hong Chen (SM'07) received her bachelor's (1992) and master's (1995) degrees, both in electrical engineering, from Southeast University, China, and her Ph.D. (2002) degree from University of Waterloo, Canada. From 1995 to 1998, she was with Nanjing Automation Research Institute (NARI), China, where she was engaged in the R&D of EMS power system applications. From 2003 to 2007, she was with ISO New England, as a principal analyst working on energy and ancillary service market design, development, and related analysis. She joined PJM Interconnection in 2007, as a senior consultant working on market and system operation. Dr. Chen is the chair of IEEE PES Power System Operation, Planning and Economics committee, and editor of *IEEE Transactions on Power Systems*, *IEEE Transactions on Smart Grid*, and *IEEE Power Engineering Letters*.

Kwok W. Cheung received his Ph.D. from Rensselaer Polytechnic Institute, Troy, NY, USA, his M.S. from the University of Texas at Arlington, Arlington, TX, USA, and B.S. from National Cheng Kung University, Taiwan, all in Electrical Engineering. Dr. Cheung has over 26 years of experience in the electric power industry. He held various technical lead and project management positions responsible for the design and implementation of a few leading energy and transmission markets worldwide. He is currently a Principal Software Architect at GE Grid Solutions (formerly Alstom Grid). Cheung has authored and co-authored over 90 technical papers published in international journals and conference proceedings and two book chapters. He is a co-holder of six US patents on power system applications. Cheung is a registered professional engineer of the State of Washington, a certified Project Management Professional of PMI and a Distinguished Lecturer of the IEEE Power & Energy Society. Dr. Cheung is a Fellow of the IEEE.

Antonio J. Conejo, professor at the Ohio State University, Columbus, OH, USA, received B.S. from Universidad P. Comillas, Madrid, Spain, M.S. from MIT, Cambridge, MA, USA, and Ph.D. from the Royal Institute of Technology, Stockholm, Sweden. He has published over 165 papers in SCI journals and is the author or co-author of books published by Springer, John Wiley, McGraw-Hill, and CRC. He has been the principal investigator of many research projects financed by public agencies and the power industry and has supervised 19 Ph.D. theses. He is an IEEE Fellow.

Pengwei Du received his B.S.E.E. and M.S. from Southeast University, Nanjing, China, in 1997 and 2000, respectively, and his Ph.D. degree in electric power engineering from Rensselaer Polytechnic Institute, Troy, NY, USA in 2006. Dr. Du is a senior engineer with the Electric Reliability Council of Texas. Prior to this, he was a senior research engineer with Pacific Northwest National Laboratory (PNNL) from 2008 to 2013.

Pavel V. Etingov (M'05) was born in 1976 in Irkutsk, Russia. He graduated with honors from Irkutsk State Technical University, specializing in electrical engineering, in 1997. He was a fellow at the Swiss Federal Institute of Technology in 2000–2001. Etingov received his Ph.D. degree in 2003 from the Energy Systems Institute of the Russian Academy of Sciences, Irkutsk, Russia. He is currently a senior research engineer at Pacific Northwest National Laboratory (PNNL), Richland, WA, USA. He is a member of the IEEE Power & Energy Society (PES), CIGRE, WECC Joint Synchronized Information Subcommittee (JSIS), WECC Modeling and Validation Work Group (MVWG), and North American SynchroPhasor Initiative (NASPI) research analysis task team. His research interests include stability analysis of electric power systems, power system operation, modeling and control, phasor measurement units (PMUs) application, wind and solar power generation, application of artificial intelligence to power systems, and software development.

Marija D. Ilić received her Doctor of Science Degree in Systems Science at Washington University in St. Louis, MO, USA in 1980. She is currently a Professor of Electrical and Computer Engineering and Engineering at Carnegie Mellon University, Pittsburgh, PA, USA, and an Affiliate Professor in the Engineering and Public Policy Department. She is the Director of the Electric Energy Systems Group (EESG) at Carnegie Mellon. She was an Assistant Professor at Cornell University, Ithaca, NY, USA, and tenured Associate Professor at the University of Illinois at Urbana–Champaign, IL, USA. She was then a Senior Research Scientist in the Department of Electrical Engineering and Computer Science, Massachusetts Institute of Technology, Cambridge, MA, USA from 1987 to 2002. She has over 30 years of experience in teaching and research in the area of electrical power system modeling and control. Her main interest is in the systems aspects of operations, planning, and economics of the electric power industry. She has co-authored and co-edited a number of books in her field of interest. Her most recent book is *Engineering IT-Enabled Sustainable Electricity Services: The Tale of Two Low-Cost Green Azores Islands*. Professor Ilić is an IEEE Fellow.

Ruiwei Jiang received B.S. degree in Industrial Engineering from the Tsinghua University, Beijing, China, in 2009, and Ph.D. degree in Industrial and Systems Engineering from the University of Florida, Gainesville, FL, USA, in 2013. Presently, he is an Assistant Professor with the Department of Industrial and Operations Engineering at the University of Michigan, Ann Arbor, MI, USA. His research interests include power system planning and operations, renewable energy management, and water distribution operations and system analysis.

Jianwei Liu (SM'07) received his bachelor's (1992) and master's (1997) from Southeast University, Nanjing, China, and Ph.D. (2004) from University of Waterloo, Waterloo, Ontario, Canada, all in electrical engineering. He also holds an MBA degree (2009) from Pennsylvania State University, USA. From 1992 to 1999, he worked in the Chinese power industry as EMS engineer and energy project manager, including the first IPP in China. From 2004 to 2007, Dr. Liu was a lead EMS specialist at ISO New England, USA. In September 2007, he joined PJM Interconnection, working on operation support and infrastructure project integration. He is now a Senior Lead Engineer leading the implementation of more than 9000 MW new generation resources and hundreds of bulk transmission upgrade projects. Dr. Liu is an active volunteer in IEEE PES and SA activities, as Utility Forum chair and task force chair. His research interests include sustainable energy system development, distributed generation and energy storage, and network security monitoring and control.

Dr. Yuri V. Makarov received his M.Sc. degree in Computers and Ph.D. in Electrical Engineering from St. Petersburg State Technical University, Russia. He was an Associate Professor at the University, conducted research at the University of Newcastle, University of Sydney, Australia, and Howard University, Washington, DC, USA. After that, he worked for Southern Company, Alabama, and occupied a position at the California Independent System Operator, California. Currently he is appointed as a Chief Scientist of Power Systems at the Pacific Northwest National Laboratory.

Alex D. Papalexopoulos (M'80–SM'85–F'01) received the Electrical and Engineering Diploma from the National Technical University of Athens, Greece, and M.S. and Ph.D. degrees in Electrical Engineering from the Georgia Institute of Technology, Atlanta, GA, USA. He is president and founder of ECCO International, a specialized energy consulting company which provides consulting and software services on electricity market design and system operations and planning within and outside the United States to a wide range of clients such as regulators, governments, ISOs/TSOs, utilities, and other market participants. He has designed some of the most complex energy markets in the world including North and South America, Western and Eastern Europe and Asia. Prior to forming ECCO International, he was a Director of the Electric Industry Restructuring Group at the Pacific Gas and Electric Company in San Francisco, California. He has made substantial contributions in the areas of network grid optimization and pricing, energy market design and competitive bidding, and implementation of EMS applications and real-time control functions. He has published numerous scientific papers in IEEE and other journals. Dr. Papalexopoulos is a Fellow of IEEE, the 1992 recipient of PG&E's Wall of Fame Award, and the 1996 recipient of IEEE's PES Prize Paper Award. He is the 2016 recipient of an honorary doctorate from the School of Electrical and Computer Engineering of the University of Patras, Greece. He is also President, CEO, and Chairman of the Board of ColorPower, a startup clean tech company focused on research, development, and commercialization of demand-side management technologies.

Mohammad Shahidehpour is the Bodine Chair Professor in the Electrical and Computer Engineering (ECE) Department and Director of Robert W. Galvin Center for Electricity Innovation at Illinois Institute of Technology (IIT), Chicago, IL, USA. He is the 2009 recipient of the honorary doctorate from the Polytechnic University of Bucharest and a Research Professor at King Abdulaziz University (Jeddah), North China Electric Power University (Beijing), and the Sharif University of Technology (Tehran). Dr. Shahidehpour was a member of the United Nations Commission on Microgrid Studies and an IEEE Fellow for his contributions to optimal generation unit commitment algorithms in electric power systems.

Chin Yen Tee is a Ph.D. candidate in the Department of Engineering and Public Policy at Carnegie Mellon University, PA, USA. She received her BA in Engineering and Economics from Smith College, Massachusetts in 2011. Her research interests include business models, regulation, and market design for the future electricity grid.

Dr. Jianhui Wang is the Section Manager for Advanced Power Grid Modeling at Argonne National Laboratory. He is the Secretary of the IEEE Power & Energy Society (PES) Power System Operations Committee. He has authored/co-authored more than 150 journal and conference publications. He is an editor of *Journal of Energy Engineering* and *Applied Energy*. He received the IEEE Chicago Section 2012 Outstanding Young Engineer Award and is an Affiliate Professor at Auburn University and an Adjunct Professor at University of Notre Dame. He has also held visiting positions in Europe, Australia, and Hong Kong including a VELUX Visiting Professorship at the Technical University of Denmark (DTU). Dr. Wang is the Editor-in-Chief of the *IEEE Transactions on Smart Grid* and an IEEE PES Distinguished Lecturer. He is the recipient of the IEEE PES Power System Operation Committee Prize Paper Award in 2015.

Jun Wu joined Alstom Grid in 2008. She is currently a senior power system engineer at GE Grid Solutions (formerly Alstom Grid). Before joining Alstom, she worked as a power system engineer in the PSASP (Power System Analysis Software Package) group at China Electric Power Research Institute (CEPRI) from 1996 to 2002. She received her B.S. from South China University of Technology, Guangzhou, China in 1990 and M.S. from CEPRI in 1996. In 2007, she studied at California State University, East Bay, CA, USA for her MBA. Jun's interests include power system analysis, market applications, and optimization. Jun is a Senior Member of the IEEE.

Lei Wu received B.S. degree in electrical engineering and M.S. degree in systems engineering from Xi'an Jiaotong University, Xi'an, China, in 2001 and 2004, respectively, and the Ph.D. degree in electrical engineering from the Illinois Institute of Technology, Chicago, IL, USA, in 2008. He was a Senior Research Associate at the Robert W. Galvin Center for Electricity Innovation at Illinois Institute of Technology from 2008 to 2010. Presently, he is an Associate Professor in the Electrical and Computer Engineering Department at Clarkson University, Potsdam, NY, USA. His research interests include power systems optimization and economics. He received the NSF Faculty Early Career Development (CAREER) Award in 2013, and IBM Smarter Planet Faculty Innovation Award in 2012. He is an IEEE Senior Member. He is an Editor of the *IEEE Transactions on Sustainable Energy* and the *IEEE Transactions on Power Systems*.

PART I

INTEGRATED SYSTEM AND MARKET OPERATION

CHAPTER 1

BALANCE ECONOMIC EFFICIENCY AND OPERATION RISK MITIGATION

Hong Chen and Jianwei Liu

SYSTEM OPERATION AND MARKET operation are tightly coupled. Electricity market operation is built upon secure system operation, trying to use market signals to address system operation needs and achieve economic efficiency. By responding to market price signals, market participants help with system operation. Therefore, the integrated system and market operation can be viewed as an engineering control system with dynamics and stability issues.

The integrated operation has a multifaceted nature. The ultimate goal is to reach the equilibrium of economic efficiency and operation risk mitigation. Finding and approximating equilibrium is an emerging frontline topic in the electricity market business.

This chapter reviews the state-of-the-art wholesale market structures and products, with the focus on their interactive impacts on daily system operations. Current challenges in approximating the equilibrium at independent system operator (ISO)/regional transmission operator (RTO) are also discussed.

Heuristic engineering efforts to approximate and achieve electricity market equilibrium at ISO/RTO have gained extensive attention from both market participants and regulatory agencies. Pennsylvania–New Jersey–Maryland (PJM)'s experience on evaluating and improving economic efficiency is discussed as a successful industrial practice in this domain. The practice of perfect dispatch (PD) at PJM has effectively measured economic efficiency in the PJM wholesale electricity market and has successfully provided guidance to system operators through daily operation. The PD practice has demonstrated over $1 billion in production cost saving in the past 8 years, a good example of the huge potential in the research domain of this book.

Power Grid Operation in a Market Environment: Economic Efficiency and Risk Mitigation, First Edition.
Edited by Hong Chen.
© 2017 by The Institute of Electrical and Electronics Engineers, Inc. Published 2017 by John Wiley & Sons, Inc.

1.1 POWER SYSTEM OPERATION RISK MITIGATION: THE PHYSICS

1.1.1 An Overview of Power System

The major components of power system are generation resources, demand resources, or load, connected by transmission facilities and distribution facilities. Power system is considered as the largest machine (or control system) in the world [1].

Generation resources can be divided based on fuel types, such as nuclear, hydro, coal, oil, natural gas, diesel, wind, and solar. For demand, normally they are not very controllable to system operation. With smart grid technologies, some are now more responsive to system conditions, called demand response. Transmission facilities include transmission lines, transformers, capacitors, reactors, phase shifters, and FACTs devices, such as static var compensator (SVC) and TCSC. Transmission facilities normally connect to the higher voltage levels, for example, 1000, 765, 500, 345, 230, 138, and 115 kV for bulk power transfer. Distribution facilities normally operate under lower voltage levels (e.g., below 115 kV). Distribution facilities bring electricity down to end customers.

Power system operation is guided by the basic circuit theory: Ohm's law and Kirchhoff's laws:

- All the injections into a node are summed to be zero.
- The distribution of the flow is based on the resistances and reactances of the branches.

All facilities have physical limitations. As a control system, power system also has its dynamic characteristics and limitations.

Power systems are normally interconnected to reduce total generation requirement, reduce total production cost, and enhance reliability. For example, in North America, there are four major interconnections: the Eastern Interconnection, the Western Interconnection, the Electric Reliability Council of Texas (ERCOT) Interconnection, and the Hydro-Quebec Interconnection [2]. In Europe, there is the synchronous grid of Continental Europe, known as European Network for Transmission System Operators for Electricity (ENTSO-E) [3]. It is the largest synchronous grid in the world.

Frequency and voltage are the two most important parameters of an interconnected power system. They have to be maintained at normal values for stable system and safety of the equipment. For example, 60 Hz frequency is operated in North America and 50 Hz system is dominant in Europe, Asia, and other parts of the world.

1.1.2 System Operation Risk Mitigation

1.1.2.1 Keep Power Balance

Electricity demand is constantly changing in the system, every hour, every minute, and every second. It is significantly impacted by weather conditions and pattern of human activities. Due to limited energy storage devices, generation has to be balanced with demand at all times, which is a moving target.

If the total generation in the system is not balanced with the total system demand, system frequency changes. Over- and under-generation can impact system frequency, causing time error. If generation is higher than demand, frequency becomes higher; if generation is less than demand, frequency becomes lower.

For interconnected power systems, the interchanges with neighboring systems are also important components in keeping power balance. Some of the transactions can be scheduled ahead of time based on the specified rules. Therefore, power balance equation can be expressed by equation (1.1):

$$\text{Total generation} = \text{total demand} + \text{total loss} + \text{net interchange} \qquad (1.1)$$

where total loss is the energy lost in the system equipment and net interchange is the net flow out of the interconnected system.

All generation resources have their physical limitations, such as time to start, minimum run time, minimum down time, minimum and maximum output, ramp rate, turnaround time, and mill points. To balance generation with demand and maintain system frequency, some generation (normally slow-start generation) has to be scheduled way ahead of time based on forecasted load. As the time is close to real time, more generation (normally fast start) is committed to balance demand. Every 5 min, generation is moved up or down to follow the load. For certain types of generating units which can move up and down within 4 s, called as regulation units, their output can be adjusted based on automatic generation control (AGC), which is often referred as secondary frequency control. The governor control of generators is often called as primary frequency control. In summary, generation is staged to balance with load and maintain system frequency.

Demand forecast, often called as load forecast, is important to schedule and dispatch generation. When scheduling generation 1 day to 1 week ahead, load is normally forecasted hourly for 24 or 168 h ahead of time. Many factors can impact load, therefore, they are factored into load forecast. The main impacting factors are temperature, humidity, wind speed, cloud covering, special social events, such as holidays or weekends. When dispatching generation in real time, very short term load forecast is used to forecast load every 5 min for 1–3 h ahead.

In North America, area control error (ACE) is used to identify the imbalance between generation and load (including interchange). Balance is measured by the frequency of the system. ACE is measured based on equation (1.2):

$$\text{ACE} = [\text{NI}_A - \text{NI}_S] - [(10 \times B) \times (F_A - F_S)] - I_{ME} \qquad (1.2)$$

where NI_A represents the actual net interchange, NI_S represents the scheduled net interchange, B represents the frequency bias constant, which is an estimate of system frequency response, F_A represents the actual frequency, and F_S represents the scheduled frequency. I_{ME} represents the interchange metering error [4].

There are variabilities and uncertainties in both generation and load. Tripped generators and sudden load increases cause the frequency to spike low while sudden large load decreases cause the frequency to spike high. To mitigate the associated power imbalance, reserves are needed in the system to control normal frequency deviation and to survive large disturbances. Reserves are the flexible unused available real power response capacity hold to ensure continuous match between generation and load during normal conditions and effective response to sudden system changes, such as loss of generation and sudden load changes. They are critical to maintain system reliability.

Reserves are secured across multiple timescales to respond to different events. The terminologies and rules vary in different systems, but they all share some fundamental characteristics. In general, some reserve types are for nonevent continuous needs; and others are for contingency events (e.g., loss of generator or facility tripping) or longer timescale events (e.g., load ramps and forecast errors). They are further categorized based on response time, online/offline status, and physical capabilities.

In North America, according to North America Electric Reliability Corporation (NERC), operating reserves are defined as "that capability above firm system demand required to provide for regulation, load forecast error, equipment forced and scheduled outages and local area protection. It consists of spinning and non-spinning reserve" [2]. Reserves are often categorized as 30 min supplemental reserve, 10 min non-spinning reserve, 10 min spinning reserve, regulating reserve, and so on. Often, regulating reserves are procured in both upward and downward directions to respond to normal load changes. They are the reserves responsive to AGC command and only carried in regulating units. Contingency reserves are used for the loss of supply, for example, generation losses. Spinning or synchronized reserves are unused synchronized capacity and interruptible load which is automatically controlled and can be available within a set period of time. Non-spinning or non-synchronized reserves are real power capability not currently connected to the system but can be available within a specified time period, which may vary in different systems.

In Europe, reserves are generally defined in three categories: primary, secondary, and tertiary control [3]. Primary control is activated within 30 s to respond to frequency deviation. Secondary control must be operational within 15 min to respond to contingency event and consists of both AGC units and fast start units. Tertiary control has a slower response to restore primary and secondary control units back to the reserve state.

The reserve requirements are also set differently in different systems. Common practices are based on the largest contingency of the system. NERC BAL-002 standard requirement is to maintain at least enough contingency reserve to cover the most severe single contingency [2]. Each region/system has different operation practices. For example, in New York system, 10-min spinning reserve requirement is set as one-half of the largest single contingency [5], while PJM's synchronized reserve requirement is set as the largest single contingency [6]. For regulating reserve, NERC does not impose explicit requirement, just to maintain sufficient regulating reserves to meet

NERC Control Performance Standards (CPS1,CPS2, and BAAL) [2]. In Europe, primary control reserves are required based on members' share of network use for energy production. Secondary control reserves are required in proportion to the maximum of yearly load in the region. Tertiary control reserve requirements are set by the individual countries [7].

With increasing penetration level of intermittent renewable resources, the reserve requirements are being reevaluated and adjusted to account for increased variability. For example, in ERCOT, forecasted wind output is factored in setting the regulating and contingency reserve requirements [8].

Interchange uncertainty poses another challenge to maintain power balance. It is volatile, hard to forecast, and significantly impacted by the market dynamics. Efforts have been started to forecast interchange in some systems, for example, PJM system.

1.1.2.2 Maintain Network Security
Network (transmission and distribution) has limited capability to transfer power from generation to load due to facility thermal, stability, and/or voltage limits. Power transfer can be restricted to any of these limits, or combination. Network security constraints are nonlinear, especially stability limits and voltage limits. Security-constrained optimal power flow (OPF) is a fundamental tool to ensure a secure operation.

1.1.2.2.1 Facility Thermal Limitation   Network facilities, such as transmission lines, transformers, have thermal ratings limiting the amount of current or apparent power that can be carried. Exceeding the thermal limits of transmission lines can cause the conductors to sag and stretch due to overheating, which could further result in faults or fires. Most equipment can be safely overloaded in certain degree. The key is how great the overload is and how long it does last. Typically, thermal ratings are set to allow specified overload for a specified period of time.

Due to thermal capabilities, the flow on any facility has to be within its thermal limit. Due to the uncertainty of facility tripping, it could overload other facilities. Therefore, the system has to be operated in a manner that it will stay within its limit under normal system condition and also under the conditions that another facility trips. The historical practice is N–1 contingency criteria, which means that when a facility trips, it will not incur overload on another facility.

In North America, normal continuous rating and emergency ratings (long term and short term) are specified for each facility [9]. Some systems have load dump ratings as well [10, 11]. Ambient temperature can affect facility thermal ratings significantly. Some systems have the thermal ratings corresponding to different temperature sets, such as PJM [10]. Dynamic line ratings are being implemented or investigated in many systems [3, 8]. The severity of thermal limit exceeding often determines corrective actions and time to correct with load shedding [10].

Power flow analysis and contingency analysis are used to determine the flow and contingency flow on the facilities. The actual flow on the facilities often comes from state estimation.

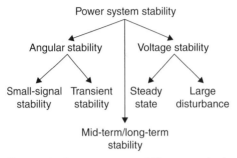

Figure 1.1   Power system stability categorization. Adapted from Kundur [12].

1.1.2.2.2 System Stability Limitation   As a control system, power system is also subject to stability limitations, that is, system should be able to return to the stable state after a disturbance. As shown in Figure 1.1, there are two main stability categories experienced in a power system, namely, angular stability and voltage stability. Each category can be further divided based on how big the disturbance is: small perturbation and large disturbance. According to [12], there is also mid-term/long-term stability which involves large voltage and frequency shift.

Voltage is the key to the overall stability of a power system. Angular stability is related to the angular separation between points in the power system; and voltage stability is related to the magnitude of the system voltages and reactive power reserves. Often, angular and voltage instability go hand in hand.

A power system is composed of many synchronous machines. Angular stability has to be maintained for the synchronization of the grid, to ensure that system torque and power angle remain controllable. The angles change as system conditions change. Interconnected power system loses synchronization when power transfer rises to extremely large magnitudes and power angles reach excessive values. Following a disturbance, transient stability becomes the concern. Power system may become instable for a period of time: angles may reach high magnitudes and rapidly change over a wide range. Synchronous generators are critical to the transient stability analysis. When torque/power angles are too large, and disturbances occur, magnetic bonds of generators may be lost. System becomes angle unstable when system operators lose their ability to control angles and power flows.

Stability analysis is often used to determine stability limits. Many power systems restrict their real power transfers due to transient stability concerns, for example, those are the power systems with long transmission lines and remote generation.

Voltage stability is the ability of a power system to maintain adequate voltage magnitudes so that when the load is increased, the power delivered to that load also increases. In a voltage-stable system, both power and voltage are controllable. Voltage stability is mainly a function of power system load. Excessive loading in the power system leads to deficiencies in reactive power and the system is no longer able to support voltage. A voltage collapse could then occur. The shortage of reactive power drives voltage instability.

When a power system experiences a voltage collapse, system voltages decay to a level from which they are unable to recover. Voltage collapse is a process during which voltage instability leads to loss of voltage in a part of the power system. A system enters a period of voltage instability prior to a voltage collapse. The effects of a voltage collapse are more serious than those of a typical low-voltage scenario. As a consequence of voltage collapse, entire systems may experience blackout. Restoration procedures are then required to restore the power system.

As power systems are pushed to transfer more and more power, the likelihood of voltage collapse occurring becomes greater. Voltage stability is mainly a concern in heavily loaded systems. Voltage stability has been responsible for major network collapse in recent years [13]. Often, system transfer capabilities are limited by steady-state voltage stability limits.

1.1.2.2.3 Voltage Limitation   All equipments are designed to operate at certain rated supply voltages. Large deviation could cause damage to system equipment. High voltages can lead to the breakdown of equipment insulation, cause transformer overexcitation, and adversely affect customer equipments. Low voltages can impact power system equipment and operations in numerous ways.

Voltage control is closely related to the availability of reactive power. The amount of available reactive support often determines power transfer limit. Heavy power transfers are a principal cause of low voltage due to the reactive power losses. Lightly loaded transmission lines are a principal cause of high voltage. Capacitors, reactors, load tap changers (LTCs), and SVCs are the equipment to control system voltage. For example, reactive support from capacitor is often needed to help prevent low-voltage problem. In system operation, reactive reserves need to be maintained and voltage deviations need to be controlled.

Often, reactive transfer interfaces are defined across the transmission paths to prevent voltage criteria violation and voltage collapse. The interface limits are used to limit the total flow over the interfaces. The reactive limits are either pre-contingency active power limits, or post-contingency active power limits. PV curves are often used to determine reactive interface limits.

1.1.2.3 Energy Management System

Energy management system (EMS) is an important tool to assist power system operation. SCADA collects measurements for system components and alarms correspondingly based on measurement. State estimation provides current system status: topology, generation, load, and power flow. State estimation relies increasingly on new technologies, such as phasor measurement units (PMUs). Network applications, such as power flow analysis, contingency analysis, voltage stability analysis, and transient stability analysis, evaluate pre- and post-contingency thermal limits, voltage limits, and stability limits. N–1 contingency rule is commonly applied in practice. Contingency element could be generation or network facilities.

When a facility overloads, directed actions, such as adjusting phase shift regulators (PARs), switching reactive devices in/out of services or adjusting generator reactive output, switching facilities in/out of services, adjusting generation of real power output via re-dispatch, and adjusting imports/exports, can be used pre-contingency to control post-contingency operation. If directed actions do not relieve an actual or simulated post-contingency violation, then emergency procedures may be directed, including dropping or reducing load as required. Thermal and voltage constraints are often controlled cost-effectively on a pre-contingency basis.

EMS is also used to perform outage analysis to evaluate if an outage has reliability impacts. Long-term analysis could be 1–6 months ahead. Short-term analysis can be 1 day, 3 days to 1 week ahead. If an outage could jeopardize system reliability, it will be canceled or rescheduled. Outage analysis directly mitigates system operation risk.

1.1.3 New Trends of Power System Operation

Power systems are under significant changes in the twenty-first century globally, with the goals of improving efficiency of electricity production, transmission, and consumption. Rapid technology innovations such as smart grid technologies, new transmission and power electronic devices, and high-efficiency energy consumption technologies are emerging and have been utilized in today's power system planning and operation.

New trends in sustainable energy system development are also observed worldwide. Environmental issues, from pollution control of fossil fuel power generation and reliability enhancement of nuclear generation facilities, to massive integration of renewable energy resources, have brought in deep social and economic impact in today's system planning and operation practices.

Economic considerations have been coherent factors in power system planning and operation. Classic stories, such as AC/DC system competition in 1900s and the emerging of power pools, have educated many generations of power engineers on the integration of power system facilities. Growth of demands in overall energy consumption, increasing constraints of nature resources, and enhanced regional (and even global) system integration are still hot topics in the planning fields. Meanwhile, thanks to the diversified energy resources and modern power electronic technologies, microgrids have been recognized as plausible approaches to modernized power system at or near the demand side. Hence, in order to adequately depict today's power grids for planners and operators, system models are becoming more and more complicated with details of new facilities, as well as dynamics and price elasticity characteristics of demands.

Electricity markets, which have been successfully implemented in many regions, have deeply reshaped power system planning and operation, philosophically and practically. Wholesale electricity markets not only provide platforms for energy transactions in forward and real-time (RT) markets, but also play key roles in system operation. Demand responses have shown significant effects on system reliability and market efficiency. Electricity markets need to be continuously developed and evolved to adapt to significant renewable resource penetration.

With these changes as background, power system planners, operation engineers, researchers, and policy makers are faced with the need of reviewing the upfront challenges in today's industry to be prepared for the future.

1.2 INTEGRATED SYSTEM AND MARKET OPERATION: THE BASICS

Electricity market operation needs to be tightly integrated with system operation, to reinforce reliable operation of the systems through strong financial incentives and bring efficiency to system operation.

1.2.1 Integrated Operation Philosophy

The objective of electricity markets is to improve economic efficiency, while risk mitigation remains the main focus of power system operation, as discussed in Section 1.1. These two different objectives often have opposite impact on resource scheduling, dispatch, and pricing. As shown in Figure 1.2, both objectives need to be respected in the integrated system and market operation. Figure 1.2 shows the integrated operation philosophy, not representing time sequences. The ultimate goal is to achieve market equilibrium, a balance between the two opposite objectives, that is, maximize total social surplus and minimize the total cost of operation risk.

 Operating criteria drive system operation practices. They are often formulated as system operation constraints in the market operation administered by ISO/RTOs. The resulting prices and dispatch signals reflect system conditions related to the modeled constraints. By responding to prices, market participants, such as generation companies, load serving entities, distribution companies, transmission

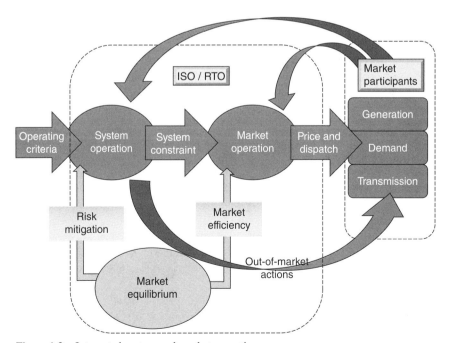

Figure 1.2 Integrated system and market operation.

companies, and financial players, help to address system operation needs. Electricity market design has been targeted toward this goal [14–18].

Ideally, the integrated system and market operation creates a closed-loop system with prices as feedback signals. Price signals align the financial interest of market participants with system and market operation objectives. Market participants are integral parts of the decision loop. By responding to price signals, they help to achieve operation objectives. Market signals, such as locational marginal prices (LMPs) and ancillary service market clearing prices, are dynamically updated to reflect changing system operation conditions. Hence, the integrated system and market operation can be viewed as an engineering control system with dynamics and stability issues [19–21].

However, due to market design and software limitation, not all operating criteria can be incorporated into current market operation. Out-of-market (OOM) actions become unavoidable. They depress market prices, increase uplift, and negatively impact economic efficiency. Therefore, OOM actions need to be reduced to improve economic efficiency.

In practical implementation, economic efficiency is often achieved by total product cost minimization or total social surplus maximization. System risk mitigation is usually modeled as power balance constraints, reserve requirement constraints, and various network security constraints in the optimization process. Attaining the balance between economic efficiency and risk mitigation has been a continual challenge with a multifaceted nature. One of the key challenges is to incorporate the cost of operation risk.

1.2.2 Current Practices

1.2.2.1 Market Design

Electricity markets are functioning around the world, from North America, shown in Figure 1.3 [22] (Pennsylvania–New Jersey–Maryland (PJM) [23], Midcontinent [24], New England [11], New York [25], California [26], ERCOT [8], Southwest Power Pool (SPP) [27], Ontario [28], Alberta [29], and New Brunswick [30]), Latin America [31] (such as Mexico [32], Chile [33], Colombia [34], Brazil [35], Argentina [36]), Europe, shown in Figure 1.4 [37] (Internal Electricity Market managed by ENTSO-E [3], including Continental Europe [38–40], Nordic in Scandinavia [41], United Kingdom [42], Ireland [43], Baltic [44–46]), to New Zealand [47], Australia [48], and some parts of Asia (e.g., Singapore [49]) and Africa [50].

Multi-settlement market design with nodal market model is generally followed in North America, with individual market variations [18]. The model details can be found in Appendices 1.A and 1.B. Figure 1.5 shows the high-level design, with the time frame ranging from planning to real time. The target is reliable and efficient RT system operation.

In the planning horizon, there are capacity markets to address resource adequacy, making sure that there will be sufficient capacity to meet future peak load plus reserve margin [11, 23–25]. Capacity markets address the "missing money"

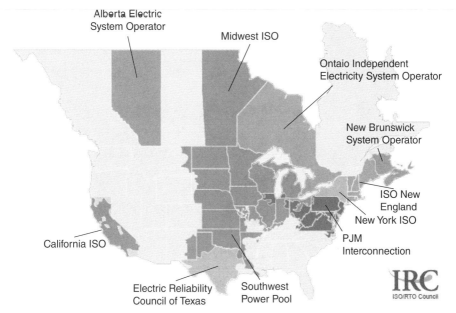

Figure 1.3   North America electricity markets. *Source*: http://www.nyiso.com/public/
markets_operations/services/planning/iso_rto/index.jsp. Reproduced with permission of the
ISO/RTO Council.

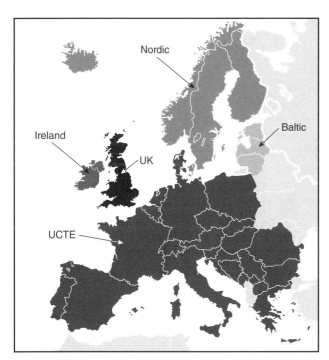

Figure 1.4   European electricity markets. *Source*: https://en.wikipedia.org/wiki/European_
Market_Coupling_Company. CC BY-SA 3.0.

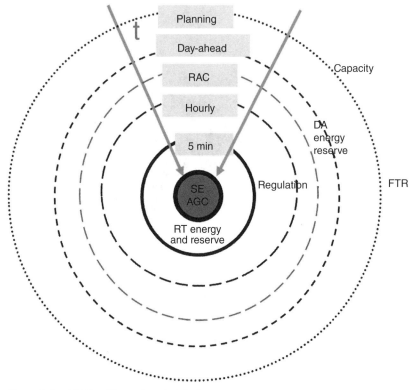

Figure 1.5   Multi-settlement market design.

issue and create long-term price signals that attract investments for both maintaining existing capacity resources and encouraging the development of new capacity resources. Locational capacity requirements are often set for capacity zones to reflect limited transfer capabilities. Some markets, for example, PJM [23] and New England [11], have long-term forward capacity markets (3 years forward). Forward design provides greater information into the reliability situation with enough lead time, so new entrants can compete against each other and avoid overbuilding. It could also lead to higher prices because generators face a bigger risk in committing to be there many years in the future. Performance-based capacity markets were recently implemented to better reward well-performing power plants and penalize those that fail to perform when needed most [11, 23]. It ensures that reliable supply will be available during extreme weather or other system emergencies. Some regions do not have capacity markets, but have resource adequacy programs to ensure long-term reliability [26, 27].

As a result of the nodal market model, prices are different at different locations to reflect transmission congestion. Financial transmission right (FTR) markets are created to hedge the risk of transmission congestion charges. FTRs are financial contracts entitling the FTR holders to a stream of revenues (or charges) based on the day ahead (DA) hourly congestion price difference across an energy path. Long-term FTR auctions, annual FTR auctions, and monthly FTR auctions are normally available for market participants to buy or sell FTRs in the central market place.

Before each operating day, DA financial markets are cleared based on bid-in demand submitted by market participants and system reserve requirements. Besides generation offers and demand bids, there are also pure financial bids, known as "virtual bids", such as INC (incremental bids), DEC (decremental bids), and up-to-congestion transactions, participating in DA markets [23]. Virtual bids help the convergence between DA markets and RT markets. The resulting DA hourly real power schedules and prices represent binding financial commitments to market participants. DA markets secure the majority of the resources for the operating day. The market clearing timelines vary for individual markets. Rules for generation offer differ in different markets as well. Generation offer information includes availability, price and/or cost offers, and operating parameters, such as ramp rates, startup time, shutdown time, minimum run time, minimum down time, and minimum and maximum generation.

Right after DA market clearing, there is re-bid period for market participants to adjust their bids. Reliability assessment commitment (RAC) processes then determine whether additional slow-start units are needed to meet forecasted load and operating reserve requirements for the operating day. In brief, RAC processes commit additional units to cover the difference between forecasted load and bid-in load, as well as the difference of corresponding operating reserve requirements. They bridge DA financial markets with physical system operation. The objectives are generally the lowest commitment cost to bring additional units online. Different objective, such as minimizing total production cost, has been discussed. In most North American markets, RAC processes are nonmarket processes with no price signal associated, except CAISO. Thus, it is an OOM action that contributes to uplift.

Within the operating day, some markets, such as PJM, New York, and California, have RT unit commitment processes to commit fast start resources with 2–3 h look-ahead time. The frequency of the case run, look-ahead time, commitment interval size, and the criteria for candidate flexible resources are different in different markets. Some markets have hourly regulation markets to secure regulation resources an hour ahead, so that the system has enough generators to provide fine-tuning that is necessary for effective system frequency control, as in the case of PJM and New England markets.

RT markets are balancing markets. Most of the RT markets co-optimize energy and reserves (such as regulation, primary reserves, contingency reserves) about every 5 min to send out dispatch and price signals, based on current system status, represented by state estimator (SE) solution, forecasted load, generators' offer information, scheduled transactions, and system topology. The resulting dispatch and price signals are sent to market participants to balance system load, maintain system reserves, and resolve system congestions. Different markets have different types of reserves co-optimized with energy in real time. The look-ahead time for RT dispatch also varies, ranging from 5 to 15 min. Longer look-ahead time tends to generate stable dispatch; shorter look-ahead time could result in higher price volatility, especially due to limited ramping capabilities. Ex-post pricing is implemented in some markets to encourage market participants to follow dispatch [14, 51].

In RT operation, AGC adjusts generation output every 4 s to fine-tune system and maintain system frequency. RT markets and system operation rely on SE, which runs periodically to capture latest system status. Some systems, such as PJM, have SE solutions every minute. Some systems have SE solutions every 3–5 min. Timely capture of the latest system conditions can help make timely RT commitment and dispatch decisions, as well as other operation decisions, therefore, better system control and risk mitigation.

As shown in Figure 1.5, when it is closer to RT, the coupling between system and market becomes much tighter. For this reason, the focus of this chapter is from DA to RT operation time frame.

In Latin America, electricity markets generally follow "hybrid" model, that is, spot energy market, to ensure economic operation and forward contract auction, called as "competition for the market" to ensure long-term resource adequacy [52].

Europe has wholesale markets for energy (DA markets and intra-day markets) and balancing and ancillary service markets for grid support services [3, 37–46]. DA markets secure most of the resources to balance demand; intra-day markets are continuous markets, critical to handle uncertainties, for example, uncertainties brought by renewable resources. To secure supply in the medium and long term, capacity mechanism has been examined and debated.

1.2.2.2 Day-Ahead and Real-Time Market Clearing

Both DA and RT market clearing are essentially bid-based security-constrained unit commitment (SCUC) and security-constrained economic dispatch (SCED) processes with nodal pricing, which is essential to the success of economically efficient electricity markets.

Figure 1.6 shows the inputs for DA market, including both physical bids and virtual bids. The system topology is based on scheduled transmission outages. DA market is a whole-day process, with 24 hourly intervals. The cleared megawatts and prices are sent to all the bidders for the corresponding hours. Economic unit scheduling, that is, SCUC/SCED, and feasibility analysis are the two key components of DA market.

For SCUC/SCED, the objective is total production cost minimization or total social surplus maximization, with power balance constraint, and branch flow limits, transfer interface limits, and other transmission-related limits as transmission security constraints. The limits of these security constraints are based on operating limits determined through system study. Reserve requirement constraints are also included to align with system reliability criteria and practices. The mathematic model of SCUC can be found in Chapter 5, Section 5.2.1.

Feasibility analysis is to ensure the physical deliverability of the 24 hourly DA schedules. It checks the network security of the economic scheduling results. Once a limit violation is identified, the corresponding constraints are then enforced into the next SCUC/SCED run. The hourly constraint sensitivities are inputted to the

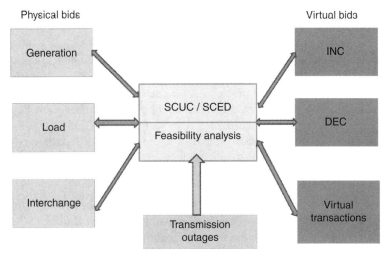

Figure 1.6   DA market clearing.

SCUC/SCED. In the case of marginal loss pricing, hourly loss sensitivities are fed into the SCUC/SCED as well.

The iteration between SCUC/SCED and feasibility analysis ends when no more limit violation is detected. The solution method details can be found in Chapter 5, Section 5.2.2.

Figure 1.7 shows the inputs for RT market. There is no virtual bid in RT market. Generation is committed and dispatched based on forecasted load and scheduled interchange for the look-ahead intervals. Some markets have multi-interval time-coupled RT solution process.

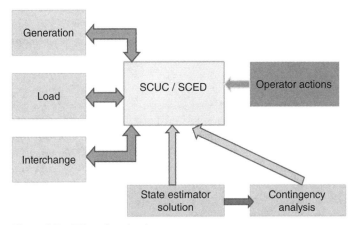

Figure 1.7   RT market clearing.

In RT market, SCUC/SCED bases on actual RT system status and topology and provides commitment and dispatch for the study intervals. The actual RT system is represented by the latest SE solution. Based on the SE solution, RT contingency analysis performs AC power flow-based contingency analysis for each defined contingency. When possible upcoming transmission limit violation is detected, system operators will investigate solutions to the problem. If generation re-dispatch is required, system operators will activate the transmission constraint to the RT SCUC/SCED. The constraint sensitivities and loss sensitivities at the current system operation point are also provided to SCUC/SCED. Binding transmission constraints indicate that generation re-dispatch is actually required to relieve the congestions. As the result of RT SCUC/SCED, the real power (MW) dispatch is generated for each resource, and the price is calculated at each pricing location.

To manage the risk posted by the uncertainties, as well as the limitation of current tools, operator actions are also important parts of the RT market operation. For example, operators can adjust load bias to reflect anticipated load forecasting error and anticipated generation changes. They can also adjust transmission limit control with certain conservativeness, as well as accounting for rapid flow increase so that desired control can be obtained early. Sometimes, operators can even manually commit units to prepare for generation bidding parameter changes and quick transmission control.

In both DA and RT markets, LMPs are generated as part of the SCED process and determined by the shadow prices of the power balance constraints and security constraints [53]. It is locational and time specific, and has three main components: energy, loss, and congestion. If there is no congestion in the system, the congestion component is zero. Under marginal loss pricing, there is loss component as well. The details can be found in Appendices 1.A and 1.B.

LMP-based congestion management is a key feature of the integrated system and market operation. Responding to LMPs, market participants could help mitigate system operation risk. LMPs encourage efficient use of the system and enhance reliability. In the long run, they encourage new generation sources to locate in areas where they will receive higher prices, signal large new users to locate where they can buy lower cost power, and encourage the construction of new transmission facilities in areas where congestion is common, in order to reduce the financial impact of congestion on electricity prices.

1.2.2.3 Intra-day Commitment/Markets

The majority of the energy/capacity is traded in DA markets. After closing of the DA market to the delivery of the next operating day, system conditions can change, sometimes, significantly: load can be significantly higher/lower than DA load; generation, for example, a big nuclear unit can trip after being cleared in DA market; natural weather condition change, such as wind or cloud covering, can cause significant renewable generation change. Intra-day markets are important for system to adjust to the changing conditions, ensuring the least cost capacity to provide power and ancillary services for the operating day.

Intra-day markets are continuous markets, critical to handle uncertainties, especially the uncertainty introduced by renewable resources. In Europe, intra-day markets happen around the clock every day, until 1 h before the delivery. It is becoming increasingly important as more wind power enters grid [54]. It is critical to the increased share of renewable energy in the energy mix. Intra-day markets result in revised flexible operation decisions to mitigate the impact of renewable uncertainties, and also allow the producers to make decisions closer to real time by adjusting intra-day bids. Overall, intra-day markets encourage efficient dispatch.

In the United States, some markets have intra-day commitment [11, 23–26]. Unit commitment is adjusted based on changing system conditions and updated offers/bids, to ensure least cost generation scheduling. However, no price signal is generated, that is, it is a nonmarket process.

1.2.2.4 Ancillary Service Markets

Ancillary services are the services necessary to support the transmission of electric power and maintain reliable operations of the interconnected system. These services generally include frequency control, spinning reserve, non-spinning reserve, replacement reserve, voltage support, and black start. Traditionally ancillary services have been provided by generators; however, with the integration of intermittent generation and the development of smart grid technologies, more resources, such as demand and batteries, can provide ancillary services now. Currently, voltage support and black start are still cost based, and there is no market yet. Ancillary service markets are therefore mainly reserve markets.

As mentioned in Section 1.1.2.1, maintaining reserves is essential in system operation. Reserve markets or ancillary service markets are created to provide market-based mechanism for the procurement of reserves on the system [55]. Transparent price signals incentivize generating and demand resources to provide flexible capability to the grid. In most US markets, reserves are jointly scheduled with energy in the DA markets and/or RT markets. Different markets have different types of reserve products co-optimized with energy [15, 51]. The detailed mathematical model can be found in Section 5.2.1.

In most of the current markets, reserve requirements are predetermined based on historical data, corresponding to operation practices, for example, N–1 or N–2 contingency criteria. To account for limited import capabilities, zonal reserve requirements are often defined, either statically or dynamically based on actual interface flow and limits. Reserve clearing prices (RCPs) are then derived from the shadow prices of the reserve requirement constraints. Reserve demand curves are usually defined for pricing reserves under scarcity [56]. In some market, post zonal reserve deployment transmission constraints are incorporated to address reserve deliverability issue [57].

To be in compliance with FERC Order 755 [58], performance-based regulation markets have been implemented in the United States. Generators are paid not only for their regulating capacity, but also for their actual performance.

To align with system operation needs, some markets are considering new reserve products, such as ramping products to procure sufficient ramping capability to handle uncertainties, as in the case of CAISO and MISO. Dynamic reserve requirements and/or dynamic reserve zones have also been extensively discussed to capture updated system requirements and accommodate new challenges, such as volatility and uncertainty brought by intermittent renewable resources and increased pressure to improve economic efficiency. However, practical implementation is still a challenge. In February 2015, FERC issued a NOPR to allow third-party provision on primary frequency response service to balancing authorities that may have a need for such a product to meet NERC Standard BAL-003-1 obligations [59, 60]. This opens the door for the market-based primary frequency response product.

1.3 ECONOMIC EFFICIENCY EVALUATION AND IMPROVEMENT: THE ECONOMICS

Increasing economic efficiency is the objective of electricity market and has been on the center of electricity market design and operation. There are many different definitions of economic efficiency. One popular and easier one is total system production cost or total social surplus as an efficiency measure. The lower the total system production cost or the higher the total social surplus, the higher the economic efficiency.

Correctly setting price leads to the most efficient use of scarce resources. Marginal cost pricing is generally applied in electricity markets, that is, setting the price of a product equal to the extra cost of producing an extra unit of output. It not only provides basis to collect revenues for suppliers, but also serves as a signal of how much the product is worth, and impacts both consumption and production level.

1.3.1 Current Practices

The general principle of electricity market design is to incorporate system needs into market signals, that is, prices, which need to be transparent and competitive. Multi-settlement market design, LMP-based congestion management, and energy and ancillary services co-optimization are consistent with this general principle.

However, as mentioned in Section 1.2.1, currently, not all system operation needs can be incorporated into market prices. Uplift or make-whole payment becomes necessary to provide incentives for resources to follow system dispatch instruction and provide production cost guarantee.

Due to the discrete nature of unit commitment problem, no exact prices can support the quantity determined in unit commitment and dispatch in the economic equilibrium [60]. The uplift caused by unit commitment is therefore unavoidable, but should be relatively small.

OOM actions are often the causes for high uplift payment, which usually indicates the inefficiency of the operation. Therefore, uplift can be used to evaluate economic efficiency [15]. Reducing uplift is one of the focuses of many electricity markets.

Considering the close coupling between system operation and market operation in real time, the main focus has been on RT uplift, which is mainly caused by RT unit commitment and other manual dispatch actions. For example, due to market modeling or software limitation, during fast load pickup time, operators may pre-position the system, such as pre-loading generation before its DA schedule, so that it comes online earlier to gain ramping capability of the system or run a unit longer than its DA scheduled hours to anticipate potential transmission problem or load increase.

To enhance economic efficiency, the primary focus has been put on incorporating system operation needs into market clearing process, so that operation risk could be mitigated through price signals. The efforts on ancillary service markets, such as explicit model ramping requirements and dynamic reserve requirements, as discussed in Section 1.2.2.4, are followed under this philosophy.

Another focus is on improving RT unit commitment, which refers to the unit commitment outside of DA market, including unit commitment during RAC process and intra-day RT operation. Using software to provide RT commitment suggestions provides more economic solution than operator manual commitment decision.

During the northeast Polar Vortex in January 2014, unprecedented increase in uplift was experienced in US northeast markets, and triggered dialogue and coordination between gas and electricity industries, as well as the design change needed for generators to update their offers to reflect gas cost volatilities. As a result, intra-day offers are being implemented in US electricity markets.

1.3.2 Perfect Dispatch at PJM

In the integrated system and market operation, one of the key challenges is to balance economic efficiency with operation risk mitigation. PJM explicitly developed the PD concept and process to address this challenge.

PD concept and process were developed to evaluate economic efficiency and improve operation, that is, manage operation risk more economically [15]. PD solution is based on actual load and generation availability, actual interchange, actual transmission topology, and actual transmission constraints, assuming that system operation criteria are given or set by external entities. ISO/RTOs need to comply with those security operation criteria. In this sense, PD solution balances economic efficiency with operation risk mitigation [15].

PD solution is obtained with all known factors, that is, without uncertainties. RT operation is, however, not perfect. There is always deviation between RT operation and PD solution. The deviation could be caused by uncertainties in load, generation, transaction or caused by dispatch actions. Deviation from RT operation to PD solution has been used to indicate how far the actual operation is from the perfect deterministic equilibrium solution, as shown in Figure 1.8 [15]. One of the deviation measurements is the difference between total RT bid-based production cost and the total bid-based production cost in PD solution, called the "perfect dispatch savings." PD savings has been used at PJM as the index to evaluate economic efficiency.

Dispatch actions have been the main focus of economic efficiency improvement, since it is the most controllable part. Dispatch actions mainly include constraint

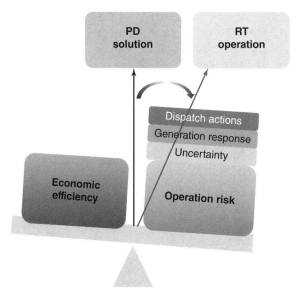

Figure 1.8   Perfect dispatch concept. *Source*: Chen and Bresler 2010 [15]. Reproduced with permission of the Institution of Engineering and Technology (IET).

control, RT steam commitment, and RT combustion turbine (CT) commitment. PD had been serving as PJM corporate goal for 7 years. The idea is to achieve economic efficiency through improving dispatch actions, which is the main contributor for RT uplift.

There were different goal focuses for different years, as shown in Figure 1.9. For the first 2 years, 2008–2009, the focus was on conservative constraint control. For 2010, the focus was on RT steam unit commitment, evaluating steam units called outside of DA Market, that is, during the RAC process. For the last 4 years, PD had been focusing on RT fast start units, mainly CT commitment, one of the major dispatch actions in RT. CT commitment, which is the process of selecting the most economic CTs to meet load and relieve congestions, is one of the most challenging dispatch actions in real time.

1.3.3   Economic Efficiency Improvement at PJM

Based on the major economic efficiency performance impacting factors, PJM has been working on improving operation practices and software tools to improve economic efficiency.

Figure 1.10 shows the PD saving due to conservative transmission limits. Diamond line shows the PD saving using the actual limiting control; and square line shows the PD saving using the 100% of the limit. For about 1 month period in 2008, the daily average PD saving difference was about $73k [15]. To improve dispatch and address conservative constraint control, PJM changed operation practice on thermal

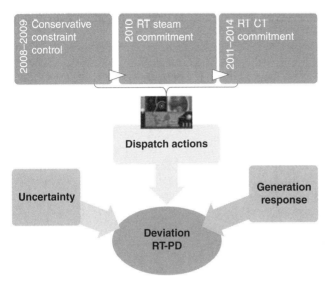

Figure 1.9   Perfect dispatch goals at PJM.

contingency constraint control starting from February 1, 2009, from 97% to 100% of emergency rating.

Constraint analysis has been provided to identify most efficient control actions for transmission constraint control, considering constraint volatility, constraint inter-actions, and unit impact patterns. Figure 1.11 shows a sample volatile constraint. Solid line represents RT flow, dotted line shows the thermal limit, and dashed line shows the controlled flow in PD solution. In RT operation, six quick-start CTs were

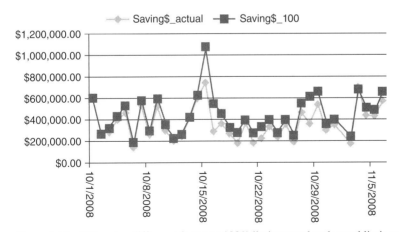

Figure 1.10   PD saving difference between 100% limit control and actual limit control. *Source*: Chen and Bresler 2010 [15]. Reproduced with permission of the Institution of Engineering and Technology (IET).

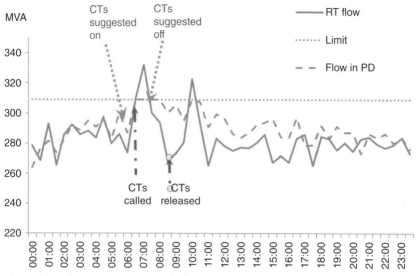

Figure 1.11   Constraint control analysis.

called before 7:00 to control the constraint when the flow was sharply increasing and approaching the limit during morning peak, shown by the dot dashed arrow. The flow quickly dropped after 7:00 when the CTs came online. The CTs were released around 9:00 after the flow dropped to 87% of the limit. After the CT release, the constraint flow bounced back and was over limit again around 10:00. With steam unit re-dispatch, the flow was then controlled under the limit. In PD solution, only two CTs were recommended to come on at 6:30 and off at 7:30, shown by the dotted arrow. The constraint flow was controlled right at the limit from 6:30 to 8:00 with both CTs and steam unit re-dispatch. Constraint analysis suggests to control earlier so that less quick-start CTs are used, which leads to overall economic solution. Controlling earlier also gives time to re-dispatch steam units for constraint control. Overcommitment often happens when the flow is over limit. More adaptive constraint control is desirable to achieve consistent control on the transmission constraints.

PJM has also been working on improving dispatch tools to help with dispatch decision making. For example, RAC application was developed to help commit slow-start units outside of DA market, embedded with three pivotal supplier (TPS) test, an integrated market mitigation process at PJM. Time-coupled multi-interval RT commitment and dispatch, IT-SCED and RT-SCED, was implemented in 2010 to provide better load pickup and peak load coverage, as well as forward-looking capability to prepare early for load or for constraint control. RT energy and reserve co-optimization was implemented in 2012 to provide market signals for reserves and overall economic solution in real time.

PD solution often suggests better RT unit commitment decision, for example, start control earlier by calling less expensive units having relatively long startup time instead of having to run expensive units later. In this case, dispatch acts more

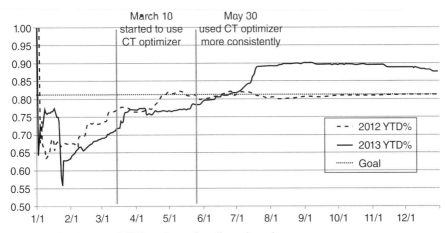

Figure 1.12   Impact of CTO to the perfect dispatch goal.

proactively, therefore, more efficiently. Based on this idea, CT optimizer (CTO) was developed to help dispatch to commit fast start units with relatively longer time-to-start (TTS) and minimum run time (MinRunTime). Previously, no tool was available for this type of units, since their parameters are outside of IT-SCED evaluation horizon, and dispatch had to make manual decision. It is often hard to make economic manual commitment. Embedded cost of high startup and no load cost and long minimum run time were often missed due to low incremental cost. It was also hard to evaluate the need of these CTs based on current system condition. CTO provides 24 h evaluation for long TTS and minimum run time fast start units, minimizing total production cost, based on latest load forecast and unit status. The results can be updated during the day when system conditions change significantly.

Figure 1.12 shows the impact of CTO to the PD performance score. Dashed line shows the 2012 year-to-date (YTD) PD score; solid line shows the 2013 PD score; and dotted line shows the 2013 PD goal. CTO was started to be used during the week of March 18, 2013. For that week, just 5 days, PD score increased 4.4%, as we can see from the sharp increase of the solid line. Starting from May 30, 2013, CTO has been used more consistently. This was one of the major factors for the significant increase in 2013 PD score. For 2013, the total PD saving was $221 million.

All these applications form well-staged unit commitment, as shown in Figure 1.13. RT operation is based on DA commitment, which sets majority of the unit commitment for the operating day. At 18:00 before operating day, based on load forecast, RAC evaluates whether additional slow-start units are needed to cover load and operating reserve for the operating day. At the beginning of the operating day, CTO provides a whole-day 24 h recommendation of fast start unit commitment. It provides a fast start unit plan for the day. Normally, dispatch uses CTO to call the fast start units which have relatively long time to start and minimum run time. CTO runs multiple times during the day, typically at the beginning of the day, or going into peak. Based on short-term load forecast (STLF), IT-SCED looks ahead 2 h to

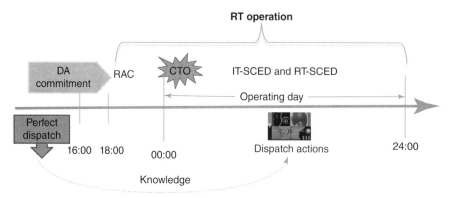

Figure 1.13 RT operation: staged unit commitment with multiple scenarios.

provide fast start unit commitment recommendation and RT-SCED looks ahead 15 min to send dispatch signals to resources. This way, slow-start resources must be committed before uncertainties are resolved and fast start resources and dispatch of all committed resources will be scheduled after uncertainties are resolved. All these tools have multiple scenarios to hedge uncertainties.

At the time of this writing, it is still challenging to incorporate uncertainties in the market clearing engines. Also, it is computationally very expensive and challenging to the stochastic or robust optimization software, especially tackling real-world problem.

From practical operation point of view, PJM addresses uncertainty through multistaged and continuous commitment. Instead of one piece of complicated software, multiple simple and fast software tools are staged for different levels of uncertainties. As we all know, when time is close to RT, uncertainty becomes much lower. For PJM system, there are lots of resources with fast startup capability, considering all uncertainties before operating day scheduling process could be overkill, especially if the gas-fired units are marginal units. Currently, over 30% of the marginal resources are gas units [23].

Dispatch makes all operation decisions. They use all these tools to operate the system. Knowledge is the key. PD provides operational analysis as knowledge for dispatch, based on previous operating days. This expert knowledge from PD as heuristic knowledge can help with setting or selecting scenarios, getting the solution needed for the system. Valuable PD analysis, commitment, and constraint control suggestions are provided to dispatch every day. The operation pattern is summarized as experiences or lessons learnt to make improvement suggestions.

In summary, staged unit commitment plus PD-based knowledge system effectively helps dispatch mitigate uncertainty and improve operation efficiency.

Figure 1.14 shows one of the ways knowledge is extracted from PD. It is a visual tool to compare fast start unit commitment at different stages, DA, CTO, IT_SCED as well as PD solution and RT. Dispatch consistently references this commitment comparison chart for historical days, and also the summary of daily operation with

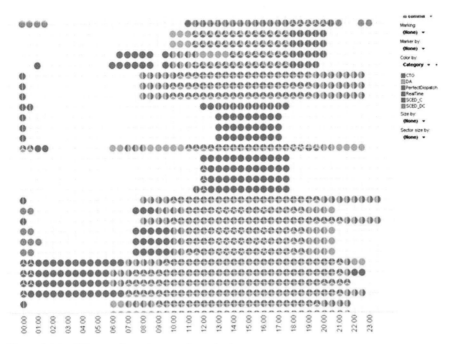

Figure 1.14   CT commitment comparison chart.

recommendations. Each color represents the commitment from one application. For example, for a unit and a specific hour, if all the applications commit it, it is shown as a five-color sliced pie.

To improve unit performance, on the one hand, unit performance impact information was provided to market participants, such as economic impact of unit performance, to increase their awareness, so that participants could provide more accurate bidding information, such as economic minimum, economic maximum, up and down ramp rate, and startup time [15]. On the other hand, efforts are being made toward accurately modeling unit characteristics, such as combined cycle unit modeling, pump storage modeling, and ramp rates, in economic scheduling and dispatch software. The idea of adaptive generation modeling has also been discussed [15].

After all these years' improvement, the economic efficiency of the system operation has been significantly improved. By the end of 2015, the cumulative production cost saving since 2008 was about $1.2 billion, as shown in Figure 1.15. RT operation becomes much closer to the perfect solution, as shown in Figure 1.16.

The effort on improving economic efficiency continues. From the beginning of 2015, PJM started to tackle uncertainty, which caused the deviation from RT to PD solution: quantifying the impact of uncertainty and further reducing the impact, as shown in Figure 1.16.

Tackling uncertainties has always been part of the system operation to mitigate operation risks. Besides load uncertainty and growing uncertainty brought by renewable resources, interchange uncertainty is becoming one of the primary

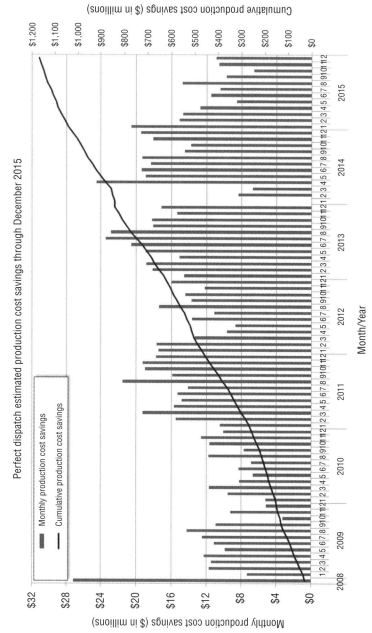

Figure 1.15 Perfect dispatch estimated production cost savings.

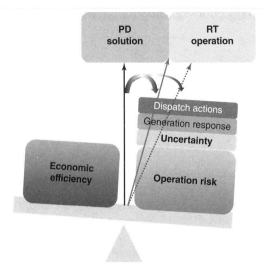

Figure 1.16   Improved economic efficiency.

uncertainty factors faced by system operators today. Uncertainties directly impact economic efficiency and remain to be a challenge. For 2015, PJM focused on tackling the load and interchange uncertainties, quantifying the monetary impact, as well as improving the processes to manage the uncertainties.

The focus was on tracking the load and interchange forecast at 18:00 of the previous operating day, used by RAC process to commit slow-start units; and the 2 h ahead load and interchange forecast used by IT-SCED to commit fast start units. The potential monetary impact was estimated based on system production cost. Each day, for each component, there was a forecast accuracy performance score calcuated based on production cost impact. The goal was to improve economic efficiency through forecast improvement.

Figure 1.17 shows the component level performance. Dot-dashed line, representing 18:00 load forecast, significantly dropped during the last 2 weeks of

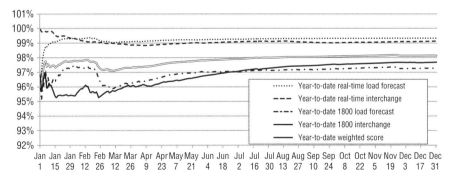

Figure 1.17   Forecast accuracy component level performance.

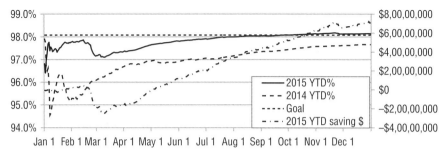

Figure 1.18   Forecast accuracy performance.

February. The new winter peak was set on February 20, 2015, at 143,295 MW. On that day, mean absolute percentage error (MAPE) of 18:00 load forecast was 1.7%, still less than 2%; however, its monetary impact was big, with low performance score of 94%. The solid line, representing 18:00 interchange, dropped first, then continuously increased from February 25, which was the day 18:00 simple interchange forecast was implemented.

Figure 1.18 shows the 2015 YTD performance compared with 2014 YTD performance. Solid line represents 2015 YTD performance; dashed line represents 2014 YTD baseline performance; dotted line represents 2015 goal; and the dot-dashed line represents the 2015 YTD monetary saving comparing to 2014. The vertical axis on the left shows the YTD performance and the vertical axis on the right shows the monetary saving. Comparing solid line with dashed line, 2015 performance was much better than 2014 baseline performance. Looking at the dot-dashed line, the YTD saving by the end of December was about $71 million. The improved efficiency comes from forecasting improvement.

As shown in Figure 1.19, to improve load forecast, multiple forecasting methods are used, from simplest similar day to neural network models with different

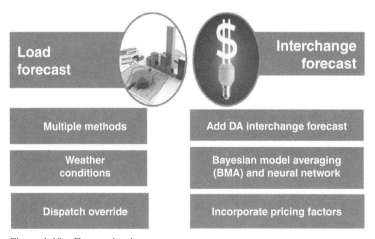

Figure 1.19   Forecasting improvement.

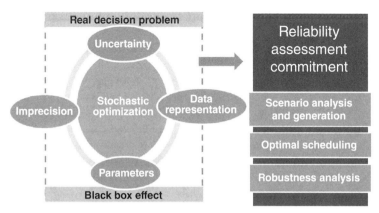

Figure 1.20   Robustness analysis of RAC.

factors considered, such as temperature and humidity. Weather condition significantly impacts system load. Better weather condition forecast, such as temperature, wind speed, humidity, and cloud covering across the region, helps to improve load forecast. Last but not least, dispatch empirical knowledge is also used. To improve interchange forecast, simple interchange forecast using similar day patterns was implemented for RAC on February 25, 2015. The accuracy has significantly improved. The performance increased from previous average of 95% to current average of 98%. Bayesian model averaging and neural network methods are also used for interchange forecast. Pricing factors are planned to be incorporated in the future to further improve interchange forecast.

In addition, to help dispatch hedge the uncertainty caused by load, RAC process is further improved to provide the corresponding load coverage for the slow-start unit commitment, based on the historical statistic data of the load. As shown in Figure 1.20, due to many uncertainty factors and data imprecision, the real decision problem is by nature a stochastic optimization problem. However, using stochastic optimization requires explicit data representation (e.g., probabilities), which is normally not known and needs strong assumptions. Further, an enormous number of parameters are introduced in the approach and can lead to black box effect, which is not desirable in the decision-aiding process. Practically, to hedge risk, robustness analysis is integrated into RAC decision. First, based on historical data, uncertainty is modeled in a manageable set of scenarios and incorporated in RAC. Then, the optimal scheduling solutions for these scenarios are obtained. The robust analysis is to evaluate the "robustness" of the scheduling solution, which is insensitive to data uncertainty and can cover a range of scenarios.

1.3.4 Challenges

1.3.4.1 DC Model in Market Clearing

DC model is generally applied in the market clearing process for the sake of solution stability and speed. However, the limit of the DC approximation, such as no losses and no reactive power calculation, brings complexity and difficulty to the modeling of losses and voltage-/reactive power-related constraints. Since RT AC power flows are available from the latest SE solution data, the information can be utilized to modify the flow calculation in the dispatch process for higher accuracy.

With DC model in the clearing engine, transmission losses are modeled as a linearization function of bus injections and withdrawals, and the total system loss from the AC solution is used to calculate the deviation from linearization. Transmission losses are explicitly modeled in power balance constraints and transmission security constraints. Load-weighted distributed slack busses are normally used as market reference to keep reference at the system load center [62].

The linear network model also limits the capability of representing reactive power-related constraints in market clearing. In terms of voltage-/reactive power-related transmission security constraints, general practice is that they are translated to the real power limits or thermal surrogate limits, then LMP-based congestion management is used to reflect the system voltage-related control needs. In the case that there is no corresponding real power limits created for voltage and voltage stability problems, operators' manual actions have to be taken, either committing units or manually dispatching down units for voltage support. These OOM actions result in uplift in market operation.

When thermal surrogate is used for voltage- and reactive power-related control, units are re-dispatched based on sensitivity of generation of real power change to the real power flow change on the thermal surrogate. Sometimes, using thermal surrogate for voltage control is not very effective and could also have undesirable system-wide impact. Meanwhile, sensitivities directly related to loading margin indicate the effectiveness of the control actions to voltage collapse problem [63–66]. It would be desirable to use loading margin sensitivities for generation re-dispatch when the system is constrained by voltage stability, so that voltage collapse could be controlled more cost-effectively. Further, load has big impact on voltage stability, and load reduction is sometimes more effective than generation re-dispatch to control voltage collapse problem. Therefore, highly nonlinear voltage stability-related constraints need to be explicitly modeled in the market clearing process [63, 64].

1.3.4.2 Deterministic Methods in Market Clearing

Currently, all the market clearing tools are based on deterministic methods. There is no easy way to incorporate uncertainty data. Stochastic programming and/or robust optimization are still not used in practical systems yet, due to modeling and computation challenges.

Under current practices, security criteria are considered set by the reliability standard. There is not much variation based on risk assessment. For example, reserve requirements are normally statically predetermined based on historical data and/or procedure and do not vary based on changing system reliability needs. Most of the times have higher level of conservativeness than required by the actual conditions, which adversely impact economic efficiency. N–1 and/or N–2 contingency criteria are also extensively used. It is easy to implement, but ignores the probabilities of contingencies. It often leads to higher cost due to over-conservativeness. Occasionally, it may adversely impact reliability.

To increase economic efficiency, considering that the probability of contingency occurrence is very small, thermal contingency control could be less conservative. The take-risk strategy could effectively reduce the operation cost [67]. The challenge is how to accurately determine the probability of the contingency, so that proper contingencies can be considered in system and market operation. At present, probability analysis has not been used in RT operation, due to data availability, model complexity, and long computing time.

Stringent deterministic methods also cause transmission constraints to bind or unbind abruptly, resulting in dispatch swing or solution volatility. Manual dispatch actions sometimes are taken under operation pressure to obtain more "stable" or "reasonable" solutions. However, manual action could adversely impact economic efficiency. With current deterministic practices, adaptive rating could help to improve economic efficiency and system reliability.

Uncertainties always exist in system operation. With current deterministic methods, it is challenging to handle the increased uncertainty brought by high penetration of renewable resources, impacting transmission constraint control, as well as properly reflecting system reserve needs. In the systems with significant percentage of renewable resources, such as California, ramping limits play critical roles in system operation. Stochastic methods will be covered in Chapters 5 and 6 to address the challenges.

1.3.4.3 *Resource Modeling in Market Clearing*
Due to software limitation, resources have limited parameters modeled in market clearing engine, normally including startup time, shutdown time, minimum run time, minimum down time, ramp-up limit, and ramp-down limit. Sometimes, the bid-in parameters cannot reflect the physical limitations, for example, dynamic ramping capabilities, startup and shutdown profiles, operating bands, turnaround time, and mill point. In this case, dispatch signals may be difficult to follow.

Combined cycle unit modeling is another challenge. The easiest way and also the most common way is to model combined cycle plants as composite units, which have same bid-in parameters as other units. With composite model, market participants have to provide proper parameters to reflect combined cycle plants' physical capabilities, which is hard to be accurate. Also, it is hard to reflect actual transmission constraint impact from each component. Some markets, such as ERCOT, use component-based model to reflect actual components' characteristics and impact. The number of combined cycle units to commit in the market is still limited in ERCOT, since most of them are self-scheduled. Therefore, performance is not a big concern there. However, for bigger markets with more combined cycle commitment in DA, performance issue definitely needs to be addressed before actual application.

Pump storage modeling is another challenge. Limited energy generation model is used for simplicity [11]. How to schedule or optimize pump storage hydro in market clearing is still a hot topic which has significant practical values. More sophisticated three-stage models, that is, pumping, generating, and offline stages, are also used

[23]. The detailed modeling of other storage devices, such as batteries, and demand responses are still needed to be explored and refined in practices.

1.3.4.4 Real-Time Commitment and Pricing

RT unit commitment has been providing incremental unit commitment based on changing system conditions. It improves economic efficiency in general.

However, due to the nature of the commitment problem, the LMPs may not justify the commitment decision, which could result in RT uplift. As an effort to further reduce uplift, different pricing mechanisms are used to reflect certain commitment cost in LMPs. Some markets have special pricing logic to allow fast start units to set price when the units are actually needed for system operation [23]; some markets amortize commitment cost in the incremental bid cost [11]; and some markets implement extended LMP model [24]. What should be the right price? There is still no definite answer for it.

Convex clearing algorithms, as those used in some places in Europe, are great in this perspective, as they produce definite prices.

1.3.4.5 Topology Flexibility

System topology is normally considered given in market clearing. In RT operation, topology control is only used for a predetermined set of transmission switching options and considered as no-cost solution. The flexibility of system topology is underused due to tool limitation. During operational planning, the outage analysis is mainly focused on reliability impact. The decision to approve or deny outage request is mainly based on reliability study. There is still no systematic way to evaluate the economic impact of outages. Some markets start to have simplified process [11, 23]. Harnessing topology flexibility is still challenging at this stage in practices and will be extensively discussed in Chapters 7 and 8.

1.3.4.6 Dispatch Time Delay and Volatility

Most of the research efforts are focused on algorithms and functional improvement. In reality, time delay is also a very important factor impacting dispatch, which is currently not explicitly factored into the algorithms yet.

As shown in Figure 1.21, from current system condition T0, to future system condition T1, total time delay can easily add up to at least 5 min currently. At PJM,

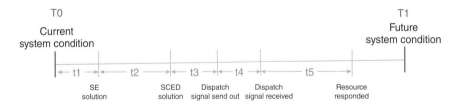

Total dispatch time delay= t1 + t2 + t3 + t4 + t5

Figure 1.21   Dispatch time delay.

the SE solution is updated every minute. In other ISO/RTOs, SE solution could be updated less frequently, for example, every 3–5 min. Therefore, t1 could be around 1 min, or longer. SCED solves based on SE solution and could take around 2 min. Dispatch evaluates SCED solutions and sends out final dispatch signals. The process t3 could take about 1–3 min. For the extreme situation, it may even take 5 min. From dispatch signal sent out, to resources receiving dispatch signal, it takes about 1 min. Resources then respond to the dispatch signal to meet future load and system needs. The time delay, t5 could take from 1 to 5 min. Therefore, the total dispatch delay can easily add up to 6–14 min. The look-ahead time, T1–T0, should be greater than the total dispatch time delay; otherwise, the solution would not be very accurate.

Due to these time delays, dispatch needs to control early to avoid potential overload. Currently, there is no good way to systematically predict how early it should be, heavily depending on dispatch experiences. Furthermore, generation response is not quite predictable either. Properly factoring time delay of price control can help improve economic efficiency of RT system operation.

1.4 FINAL REMARKS

Figure 1.22 summarizes the road map toward better price signals to reflect operation risk mitigation and incentivize resources for more flexibility to respond to system operation needs, therefore, improved economic efficiency. It captures latest industry practices, which is still changing.

There are three main stages of market design and implementation.

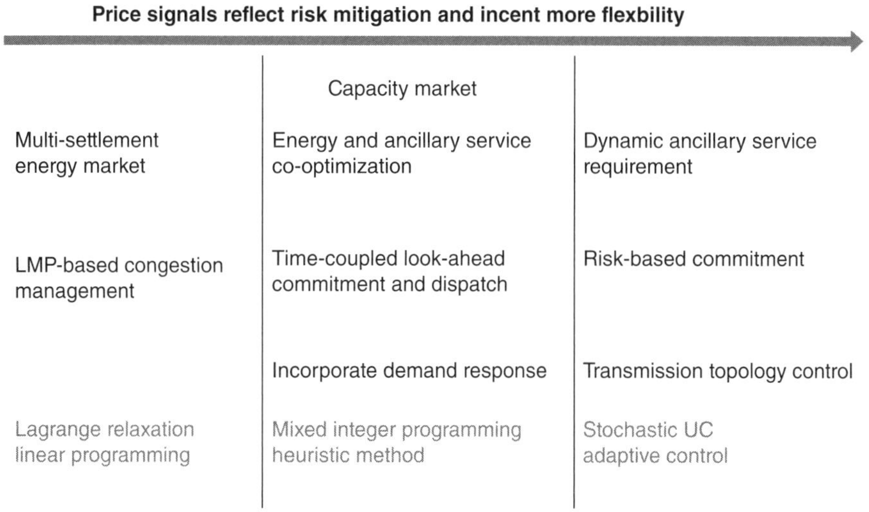

Price signals reflect risk mitigation and incent more flexbility

	Capacity market	
Multi-settlement energy market	Energy and ancillary service co-optimization	Dynamic ancillary service requirement
LMP-based congestion management	Time-coupled look-ahead commitment and dispatch	Risk-based commitment
	Incorporate demand response	Transmission topology control
Lagrange relaxation linear programming	Mixed integer programming heuristic method	Stochastic UC adaptive control

Figure 1.22   Road map.

The first stage sets the basic market design: multi-settlement energy market and LMP-based congestion management. During the first stage, solution algorithm was mainly Lagrange relaxation and linear programming.

The second stage starts to develop capacity market to provide long-term reliability price signals, implement ancillary service markets which is co-optimized with energy to provide overall economic efficiency, use time-coupled look-ahead commitment and dispatch to help RT operation, and also incorporate demand response in the market to help both reliability and efficiency. During this stage, mix-integer programming is used in the solution, and different heuristic methods are also developed to help with market solution.

Markets are further evolving into the third stage, working on dynamic ancillary service requirements and risk-based commitment to handle increased system uncertainties, and exploiting transmission topology control for both reliability and economics. These new development features need new algorithms to support, such as stochastic unit commitment and adaptive control. The rest of the book will address some of these challenges.

APPENDIX 1.A NOMENCLATURE

i	Index for all generators
j	Index for all price-sensitive demands
t	Index for all reserve types, for example, 10-min synchronized (or spinning) reserve, 10-min non-synchronized reserve, and 30-min operating reserve
k	Index for all transmission constraints
s	Index for all reserve constraints
n	Index for all busses in the system
G	Set of generators
RC	Set of reserve category
min	Superscript that indicates the minimum value
max	Superscript that indicates the maximum value
SE	Superscript that indicates values obtained from state estimator solution
10	Superscript that indicates values that can be achieved in 10 min
30	Superscript that indicate values that can be achieved in 30 min
up	Superscript that indicates the upper bound
dn	Superscript that indicates the lower bound
c	Offer price for generators
b	Bid price for price-sensitive demand
o	Offer price for reserves
R	Reserve quantity

P	Generation output
D	Output of price-sensitive demand
L	Fixed demand
flow	Real power flow over a transmission line or a transmission corridor
Q	Reserve requirement
P_{loss}	System real power losses
$\delta^x_{t,i}\text{or}j$	Binary value: 1 represents that reserve type t from generator i or price-sensitive demand j belongs to the set x
RC^T_i	Ramping capability of generator i within the dispatch interval T
RC	Ramping capability for generators or price-sensitive demand
λ	Shadow price of energy balance constraint
μ	Shadow prices of transmission security constraints
β	Shadow prices of reserve requirement constraints
γ	Shadow prices of capacity constraints
η	Shadow prices of ramp constraints, or upper/lower bounds of generators and price-sensitive demand
$\beta^{min}_{t,i}$	Shadow prices of lower bounds for reserve quantity

APPENDIX 1.B ELECTRICITY MARKET MODEL

Most of the electricity markets are administered by ISO/RTO or market administrators. Market participants include generation companies (GenCo), distribution companies (DisCo), load serving entities (LSE), transmission companies (TransCo), pure financial companies, large industry load, or load aggregators.

Offers are submitted for generators and bids are submitted for dispatchable demand. Uniform marginal pricing is commonly used in current electricity market clearing, that is, market price is determined by the intersection of supply curve and demand curve [68]. All supply units with prices lower or equal to the market price are scheduled to produce.

Energy and reserve co-optimization-based market clearing model is generally used. The objective function is to minimize the total production cost, as shown in equation (1.B.1)

$$\text{Min} \sum_{i \in G} \left(c_i P_i + \sum_{t \in \text{RC}} (o_{t,i} \cdot R_{t,i}) \right) + \sum_j \left(-b_j \cdot D_j + \sum_{t \in \text{RC}} (o_{t,j} \cdot R_{t,j}) \right) \tag{1.B.1}$$

subject to the following constraints:

(1) Energy balance constraint:

$$(\lambda) : \sum_{i \in G} P_i = \sum_n L_n + \sum_j D_j + P_{\text{loss}} \tag{1.B.2}$$

where interchanges are considered in the fixed demand, L_n in this case. System loss is linearized as following:

$$P_{\text{loss}} = \sum_i \left(\frac{\partial P_{\text{loss}}}{\partial P_i} \cdot P_i \right) + \sum_n \left(\frac{\partial P_{\text{loss}}}{\partial L_n} \cdot L_n \right) + \sum_j \left(\frac{\partial P_{\text{loss}}}{\partial D_j} \cdot D_j \right) + \text{offset}$$

where offset is the linearization error [62].

(2) Transmission security constraints:

$$(\mu_k) : \text{flow}_k \leq \text{flow}_k^{\max} \tag{1.B.3}$$

Based on linearization, transmission flow is represented as following:

$$\text{flow}_k = \sum_i \left(\frac{\partial \text{flow}_k}{\partial P_i} \cdot P_i \right) + \sum_n \left(\frac{\partial \text{flow}_k}{\partial L_n} \cdot L_n \right) + \sum_j \left(\frac{\partial \text{flow}_k}{\partial D_j} \cdot D_j \right) + \text{bias}$$

where bias is the linearization error.

(3) Reserve constraints:

$$(\beta_s) : \sum_i \sum_t \left(\delta_{t,i}^s R_{t,i} \right) + \sum_j \sum_t \left(\delta_{t,j}^s R_{t,j} \right) \geq Q_s \tag{1.B.4}$$

(4) Capacity constraints for units:

$$(\gamma_i) : P_i + \sum_t R_{t,i} \leq P_i^{\max} \tag{1.B.5}$$

(5) Capacity constraint for dispatchable demand:

$$(\gamma_j) : D_j - \sum_t R_{t,j} \geq D_j^{\min} \tag{1.B.6}$$

(6) Ramp constraints for generating units:

$$\left(\eta_i^{\text{up}} \right) : P_i \leq P_i^{\text{SE}} + \text{RC}_i^T \tag{1.B.7}$$

$$\left(\eta_i^{\text{dn}} \right) : P_i \geq P_i^{\text{SE}} - \text{RC}_i^T \tag{1.B.8}$$

(7) 10-min ramp capability constraints:

$$\left(\eta_i^{10} \right) : \sum_t \left(\delta_{t,i}^{10} \cdot R_{t,i} \right) \leq \text{RC}_i^{10} \tag{1.B.9}$$

$$\left(\eta_j^{10} \right) : \sum_t \left(\delta_{t,j}^{10} \cdot R_{t,j} \right) \leq \text{RC}_j^{10} \tag{1.B.10}$$

(8) 30-min ramp capability constraints:

$$\left(\eta_i^{30} \right) : \sum_t \left(\delta_{t,i}^{30} \cdot R_{t,i} \right) \leq \text{RC}_i^{30} \tag{1.B.11}$$

$$\left(\eta_j^{30} \right) : \sum_t \left(\delta_{t,j}^{30} \cdot R_{t,j} \right) \leq \text{RC}_j^{30} \tag{1.B.12}$$

(9) Upper and lower bounds of decision variables

$$\left(\eta_i^{\min}\right) : P_i \geq P_i^{\min} \tag{1.B.13}$$

$$\left(\eta_i^{\max}\right) : P_i \leq P_i^{\max} \tag{1.B.14}$$

$$\left(\eta_j^{\min}\right) : D_j \geq D_j^{\min} \tag{1.B.15}$$

$$\left(\eta_j^{\max}\right) : D_j \leq D_j^{\max} \tag{1.B.16}$$

$$\left(\breve{\beta}_{t,i}^{\min}\right) : R_{t,i} \geq 0 \tag{1.B.17}$$

$$\left(\breve{\beta}_{t,j}^{\min}\right) : R_{t,j} \geq 0 \tag{1.B.18}$$

For constraints (1.B.2)–(1.B.18), λ, μ, β, γ, η, and $\breve{\beta}$ are the corresponding shadow prices, that is, Lagrange multipliers.

Based on marginal pricing, the LMP for each bus n is defined by equation (1.B.19):

$$\text{LMP}_n = \lambda \left(1 + \frac{\partial P_{\text{loss}}}{\partial D_n}\right) - \sum_k \left(\mu_k \cdot \frac{\partial \text{flow}_k}{\partial D_k}\right) \tag{1.B.19}$$

The RCP is defined by equation (1.B.20):

$$\text{RCP}_{t,i} = \sum_s (\beta_s \cdot \delta_{t,i}^s) \tag{1.B.20}$$

When the capacity constraint binds, energy price LMP_n is coupled with reserve clearing price $\text{RCP}_{t,i}$.

REFERENCES

[1] A. Gomez-Exposito, A. J. Conejo, and C. Canizares, *Electric Energy Systems: Analysis and Operation*, CRC Press, 2008.

[2] www.nerc.com

[3] www.entsoe.eu

[4] "Balancing and Frequency Control," NERC, January 26, 2011.

[5] "Ancillary Services Manual," New York Independent System Operator, December 2015.

[6] "PJM Manual 13: Emergency Operations," PJM, January 1, 2016.

[7] M. Milligan, P. Donohoo, D. Lew, E. Ela, B. Kirby, H. Holttinen, E. Lannoye, D. Flynn, M. O'Malley, N. Miller, P. B. Eriksen, A. Gøttig, B. Rawn, J. Frunt, W. L. Kling, M. Gibescu, E. G. Lázaro, A. Robitaille, and I. Kamwa, "Operating Reserves and Wind Power Integration: An International Comparison." Available at http://www.nrel.gov

[8] www.ercot.com

[9] "System Operating Limit Definition and Exceedance Clarification," NERC, August 4, 2014.

[10] "PJM Manual 3: Transmission Operations," PJM, December 1, 2015.

[11] www.iso-ne.com

[12] P. Kundur, *Power System Stability and Control*, McGraw-Hill, Inc., 1994.

[13] J. Hossain and H. R. Pota, *Robust Control for Grid Voltage Stability: High Penetration of Renewable Energy*, Springer, 2014.

[14] A. Ott, "Experience with PJM market operation, system design, and implementation," *IEEE Transactions on Power Systems*, vol. 18, no. 2, pp. 528–534, May 2003.

[15] H. Chen and S. Bresler, "Practices on real-time market operation evaluation," *IET Generation, Transmission and Distribution*, vol. 4, no. 2, pp. 324–332, 2010.

[16] H. Chen and J. Liu, "Practices of System Security Pricing," IEEE/PES 2008 General Meeting, Pittsburg, USA, July 2008.

[17] H. Chen and S. Bresler, "Risk Management in PJM's Market Operation," IEEE/PES 2008 General Meeting, Pittsburg, USA, July 2008.

[18] Standard Market Design (SMD), Notice of Proposed Rulemaking (NOPR), FERC Docket No. RM01-12-0000, November 11, 2002.

[19] F. L. Alvarado and W. S. Mota, "The Role of Energy Imbalance Management on Power Market Stability," Proceedings of 31th Hawaii International Conference on System Sciences, Volume III, pp. 4–8, January 6–9, 1998.

[20] F. L. Alvarado, J. Meng, C. L. DeMarco, and W. S. Mota, "Stability analysis of interconnected power systems coupled with market dynamics," *IEEE Transactions on Power Systems*, vol. 16, no. 4, pp. 695–701, November 2001.

[21] F. L. Alvarado, "Is System Control Entirely by Price Feasible," Proceedings of the 36th Hawaii International Conference on System Sciences, 2003.

[22] www.isorto.org

[23] www.pjm.com

[24] www.misoenergy.org

[25] www.nyiso.com

[26] www.caiso.com

[27] www.spp.org

[28] www.ieso.ca

[29] www.aeso.ca

[30] tso.nbpower.com

[31] H. Rudnick, "The Electricity Restructuring in South America: Successes and Failures on Market Design," Harvard Electricity Policy Group, San Diego, CA, January 29–30, 1998.

[32] www.cenace.gob.mx

[33] www.cne.cl

[34] www.xm.com.co

[35] www.ccee.org.br

[36] www.cammesa.com

[37] www.eurelectric.org

[38] www.centrel.org/ucte.html

[39] www.eex.com/en/

[40] www.omie.es/en/inicio

[41] www.nordpoolspot.com/#/nordic/table

[42] www.ofgem.gov.uk/electricity

[43] www.sem-o.com

[44] www.elering.ee

[45] www.ast.lv

[46] www.regula.lt/elektra/Puslapiai/default.aspx

[47] www.ea.govt.nz

[48] www.aemo.com.au

[49] www.ema.gov.sg

[50] http://www.ecowapp.org/

[51] T. Zheng and E. Litvinov, "Ex post pricing in the co-optimized energy and reserve market," *IEEE Transactions on Power Systems*, vol. 21, no. 4, pp. 1528–1538, November 2006.

[52] T. J. Hammons (Ed.), "Market mechanisms and supply adequacy in the power sector in Latin America," in *Electricity Infrastructures in the Global Marketplace*, 2011, ISBN: 978-953-307-155-8.

[53] X.W. Ma, D. Sun, and K. W. Cheung, "Evolution toward standardized market design," Invited paper, *IEEE Transactions on Power Systems*, vol. 18, no. 2, pp. 460–469, 2003.

[54] http://networkcodes.entsoe.eu/category/intraday-markets

[55] J. H. Chow, R. deMello, and K. W. Cheung, "Electricity market design: an integrated approach to reliability assurance," Invited paper, *IEEE Proceeding* (Special Issue on Power Technology & Policy: Forty Years after the 1965 Blackout), vol. 93, no. 11, pp. 1956–1969, November 2005.

[56] W. Hogan, "A Model for a Zonal Operating Reserve Demand Curve," October 15, 2009. Available at http://www.whogan.com

[57] Y. Chen, P. Gribik, and J. Gardner, "Incorporating post zonal reserve deployment transmission constraints into energy and ancillary service co-optimization," *IEEE Transactions on Power Systems*, vol. 29, no. 2, pp. 537—549, March 2014.

[58] "Frequency Regulation Compensation in the Organized Wholesale Power Markets," Docket Nos. RM11-7-000 and AD10-11-000, Order No. 755, October 20, 2011.

[59] "Third-Party Provision of Primary Frequency Response Service," Docket No. RM15-2-000, February 19, 2015.

[60] "Frequency Response and Frequency Bias Setting," NERC, February 7, 2013.

[61] P. Gibrik, W. Hogan, and S. L. Pope, "Market-Clearing Electricity Prices and Energy Uplift," December 31, 2007. Available at http://ksghome.harvard.edu/-whogan

[62] E. Litvinov, T. Zheng, G. Rosenwald, and P. Shamsollahi, "Marginal loss modeling in LMP calculation," *IEEE Transactions on Power Systems*, vol. 19, no. 2, pp. 880–888, May 2004.

[63] H. Chen, "Security Cost Analysis in Electricity Markets Based on Voltage Security Criteria and Web-Based Implementation," Ph.D. Dissertation, University of Waterloo, Waterloo, ON, Canada, 2002.

[64] H. Chen, C. Canizares, and A. Singh, "Web-based security cost analysis in electricity markets," *IEEE Transactions on Power Systems*, vol. 20, no. 2, pp. 659–667, May 2005.

[65] S. Greene, I. Dobson, and F. L. Alvarado, "Sensitivity of the loading margin to voltage collapse with respect to arbitrary parameters," *IEEE Transactions on Power System*, vol. 12, no. 1, pp. 262–272, February 1997.

[66] S. Greene, I. Dobson, and F.L. Alvarado, "Sensitivity of transfer capability margins with a fast formula," *IEEE Transactions on Power System*, vol. 17, no. 1, pp. 34–40, February 2002.

[67] H. Chen, C.A. Canizares, and A. Singh, "Transaction Security Cost Analysis by Take-Risk Strategy," Proceedings of 14th Power Systems Computation Conference 2002, Spain, June 2002.

[68] R. Ethier, R. Zimmerman, T. Mount, W. Schulze, and R. Thomas, "A uniform price auction with locational price adjustments for competitive electricity markets," *Electrical Power and Energy Systems*, vol. 21, pp. 103–110, 1999.

DISCLAIMER

The views and opinions expressed in this chapter are those of the authors and do not necessarily reflect the official position of PJM.

CHAPTER 2

MITIGATE MARKET POWER TO IMPROVE MARKET EFFICIENCY

Ross Baldick

2.1 INTRODUCTION

This chapter will discuss the interaction of market power and market efficiency, and particularly consider the implications of market power mitigation in improving the efficiency of operation of electricity markets. The definition of market power adopted in this chapter is a standard one from economics: the ability to profitably alter prices away from competitive levels, typically by withholding supply so that the resulting prices are higher than they would be otherwise. An extreme version of market power is where a market participant is "pivotal;" that is, where the supply from the other market participants is insufficient to meet demand. Efficiency will be defined as the production of electricity at least overall cost.

2.1.1 Competitive Prices

Competitive price levels will be defined more carefully in Section 2.2, but, briefly, they are the prices that would prevail in a clearing-price market where no market participant can individually affect prices. The latter condition typically occurs in electricity markets in the circumstances that supply is abundant compared to desired demand and the transmission constraints are not binding. The clearing price is then the highest offer price amongst the dispatched generators. Under competitive conditions, offers can be expected to reflect marginal costs and so the clearing price is the highest marginal cost amongst the dispatched generators that are not operating at minimum or maximum. We will refer to this as the marginal cost of the marginal generator.

Centralized dispatch in electricity markets is explicitly designed to result in production of energy at least cost under competitive circumstances. That is, it results in efficient dispatch. While there are several aspects to market efficiency, efficient dispatch and efficient capital formation will be the principal aspects of market efficiency considered in this chapter.

Power Grid Operation in a Market Environment: Economic Efficiency and Risk Mitigation, First Edition.
Edited by Hong Chen.

If transmission constraints bind, the prices differentiate by bus, or nodally. Competition is at least somewhat limited when this occurs, so that market participants are more likely to be able to affect prices and therefore are more likely to have market power. Nevertheless, we can still consider idealized competitive prices in the case of binding transmission constraints. In this case, there can be several marginal generators at different buses, but competitive prices still reflect marginal costs of the marginal generators.

When supply is scarce so that all generators are fully dispatched, competitive prices are, in principle, determined by the willingness-to-pay of demand, potentially as specified in demand bids. Demand willingness-to-pay is generally much higher than typical generator offer prices. Prices under scarcity are therefore generally much higher than the typical marginal costs of generators. To summarize, competitive prices will typically reflect marginal costs of the marginal generators when there is ample supply. When supply is tight, competitive prices can be expected to be much higher, reflecting demand willingness-to-pay.

Prices based on both supply offers and demand bids under competitive conditions have incentives, under favorable assumptions, for the development of efficient levels of capital investment that lead to overall least capital and operating costs. That is, prices result in transfers from the demand side to the supply side that compensate both operations and capital expenditures and result in efficient levels of capital.

2.1.2 Market Power and Market Power Mitigation

Unfortunately, competitive conditions do not always prevail in restructured electricity markets. In part because of the legacy from pre-restructuring arrangements, typical electricity markets today have relatively little demand-side price-setting participation. In some cases, and in some markets, demand is not even able to specify a bid.[1] In the extreme, load is a fixed value and when supply is tight, many market participants on the supply side may be able to affect prices. Under these circumstances, it can be unclear as to what should be the competitive price level, complicating the remuneration of capital investment. Particularly since competitive prices under these circumstances would be high anyway, market power continues to be problematic in the operation of restructured electricity markets when supply is tight and

---

[1] Demand in current markets exhibits very little price elasticity, in part because retail tariffs typically do not reflect the full, or any, variability of wholesale markets, and in part because of the lack of ability of demand to explicitly bid into the market. However, it is unclear as to whether there would be considerably more exhibited elasticity under conditions where demand was exposed to market prices or if (representative of) demand were required to bid willingness-to-pay. If demand that was exposed to varying prices in fact remained inelastic, then arguably the exercise of market power would not lead to any "allocative inefficiency." On the other hand, if demand were actually able to respond to prices, then the exercise of market power would have the further drawback of resulting in inefficient allocation. Although price elasticity will be considered in this chapter, the discussion will primarily be in its role of making markets more competitive, rather than the issues regarding allocative efficiency.

particularly when transmission constraints bind. Consequently, electricity markets continue to need market power mitigation.

There are two main goals of such market power mitigation:

1. avoiding transfers that are significantly above competitive levels, and
2. achieving efficiency.

Much discussion of market power mitigation focuses on the first goal or even focuses simply on avoiding "large" transfers. As a vivid example of this issue, the "California electricity market crisis" of 2000 involved particularly high wholesale prices due to the exercise of market power that resulted in large transfers, but "even a perfectly competitive California electricity market would have seen wholesale electricity expenditures triple between the summers of 1998 and 2000," [1, p. 1377]. That is, even had there been competitive conditions in California in 2000, prices would still have been high because marginal costs were high and would have attracted scrutiny about high, albeit appropriate, transfers. This is due to the fact that when supply is tight, peaking generators with the highest marginal costs will be dispatched. In the California case, there were additional issues such as high natural gas prices and high emissions costs in 2000 that further increased the marginal cost of the peaking generators. With the exercise of market power, prices were even higher than the high competitive prices, and the transfers were large.

This chapter will, however, focus on the second goal of market power mitigation of achieving efficiency. Market power mitigation in practice may not achieve either of the goals of limiting transfer to competitive levels and achieving efficiency. While regulation can typically reduce the level of transfers, there may not be systematic ways to set prices at competitive levels in the absence of demand-side price setting, and the resulting market may be operationally inefficient, particularly in the context of transmission constraints.

Moreover, at least in principle, exercise of market power and excessive transfers are not always coincident with inefficiency. For example, consider a symmetric electric generation industry where each generation owner has the same distribution of types of generators in its portfolio. Exercise of market power may in this case result in high prices but dispatch may be operationally efficient.

Even if efficiency were achieved in the operational context under conditions of market power, it can be expected that the resulting high prices will result in poor capital investments. In the absence of barriers to entry, high prices may result in over-investment. In practice, typical electricity industries have market participants with heterogeneous portfolios. As discussed in Reference [1, p. 1379], the exercise of market power by one participant, with that participant generating at levels that are lower than competitive, will result in prices that are higher than competitive. In this case, other market participants may produce more than they would have under competitive conditions since the price is higher than under competitive conditions. The dispatch will typically be inefficient in this case and poor capital investments may result.

To summarize, while it is in principle possible for exercise of market power to result in efficient dispatch, more typically exercise of market power will result in inefficient dispatch and it will typically also be associated with inefficient capital formation. Achieving competitive prices will therefore be adopted in this chapter as the normative goal, requiring both that competitive prices be estimated under particular conditions and that market outcomes corresponding to competitive conditions will be enforced.

2.1.3 Measuring Competitiveness and Market Power

Estimating competitive outcomes presents challenges, particularly under scarcity. In several markets, including ERCOT, Alberta, and Australia, exercise of market power is or has been condoned in the interest of approximating the competitive pricing outcome under scarcity conditions. ERCOT's version of this rule is even called the "scarcity pricing mechanism," although it is not a "mechanism" in the formal sense of economics and does not directly measure scarcity.[2] That is, there are at least somewhat contradictory issues in place in the regulation of market power, complicating the situation. Firms that are given any dispensation to set high prices under these circumstances will typically produce less than they would under competitive conditions, resulting in the inefficient dispatch condition described above and not achieving the competitive outcome. The resulting prices are unlikely to correspond to competitive conditions, as defined in this chapter.

To understand market power in detail, it must first be defined and measured, whereas the discussion above has been qualitative. We now turn to quantitative measures of market power. Traditional approaches to market power as considered by the United States Department of Justice have derived from concerns about collusion, and typically utilize measures that reflect market concentration; that is, the level to which ownership of generation is shared amongst relatively few firms. While collusion is possible in restructured electricity markets, particularly given the repeated nature of electricity market interactions [2], unilateral exercise of market power is also relevant in the electricity industry. Particularly in the context of transmission constraints, the traditional approaches, deriving from analysis of collusion, have typically been applied in an *ad hoc* manner to situations where even unilateral exercise of market power are problematic.

This chapter will begin from a more direct measure: when total industry capacity is not fully dispatched to capacity, the measure of market power for a market

---

[2] The electricity industry, and particularly the restructured electricity industry, is replete with confusing jargon and euphemisms. As mentioned in the text, the ERCOT "scarcity pricing mechanism" is neither a *mechanism* nor does it *price scarcity*, although the more recently implemented "operating reserve demand curve" does in fact price scarcity. As another example relevant to the discussion of market power, "price takers" in the sense typically used in electricity markets are *not* necessarily price takers in the economics sense of not being able to affect prices. In the context of transmission constraints, "congestion cost" is very typically used by market participants not to refer to a cost, but rather to a rental due to transmission limitations.

participant that is also not fully dispatched to its capacity limit is the price mark-up or "margin between the [market] price and the marginal cost of the highest cost unit [that would be] necessary to meet demand" under competitive conditions [1, p. 1383]. (The cases where a market participant is fully dispatched and where total supply is at its limit are discussed in the following paragraphs.) That is, the mark-up measures the deviation of market prices from competitive prices, defined to be the marginal cost of the marginal generator under efficient dispatch that minimizes total operating costs.

A fully dispatched market participant might benefit from the exercise of market power by others, but is not exercising market power in the sense that it is not withholding, and we will define its mark-up to be zero. A market participant with multiple generators may have some fully dispatched, while it may be withholding at others. Consequently, the price paid to the fully dispatched generator may be above competitive, and we will consider a mark-up in that case.

When total supply is at its capacity limit under competitive conditions, the marginal cost of the highest cost unit is replaced in this definition by the lowest willingness-to-pay amongst the served loads. As mentioned above, this willingness-to-pay is typically much higher than the typical marginal cost of generators and, consequently, high prices under these circumstances are not indicative of the exercise of market power.

The Lerner index is another measure used in the context of market, and provides a relative measure of market power. It is equal to the mark-up divided by the price and therefore ranges from zero for competitive conditions to (in principle) one, assuming prices are non-negative. Because prices in electricity markets can vary so significantly (and can even be negative), the Lerner index can be somewhat misleading in electricity markets. For this reason, we will focus on the mark-up as a measure of market power rather than the Lerner index.

To summarize, the actual (or estimated) price is compared to the (estimated) price under competitive conditions, and the measure of market power for a market participant that is not fully dispatched is the price mark-up above the competitive price. As the careful *ex post* analysis of conditions in California from 2000 described in Reference [1] shows, however, assessing the competitive conditions can require extensive effort and simulation.

Market power will generally be greater where there is higher market concentration-that is, fewer market participants with higher market shares—other things being equal. The mark-up will change over time as demand (and supply) conditions change. The significance of the distortion from competitive prices as measured by the mark-up therefore depends not only on the size of the mark-up at a particular time, but also on the temporal variation of the mark-up.

Some approaches to measuring market power aim at *ex ante* assessment of market power, presumably necessitating some form of worst case or typical case analysis, based on simulation of prices under competitive conditions and under the conditions where market power might be exercised. In contrast, "conduct and impact" type analysis aims at assessing the effect on actual conditions, typically using actual market outcomes and the history of offers (and an estimate of competitive conditions). Conduct and impact tests will be discussed in Section 2.5.2.

2.1.4 Transmission Constraints

Transmission constraints add a layer of complexity to this assessment since they can effectively limit competition and thereby increase the "local" market concentration, as well as produce localized conditions of scarcity. Approaches to modeling and mitigating transmission-constrained market power in ISOs today tend to focus on individual constraints. For example, the PJM, CAISO, and ERCOT transmission-constrained market power analysis approaches involve attempts to assess each transmission constraint individually as being either "competitive" or "non-competitive," neglecting the interactions between multiple potentially binding constraints, where by "binding" it is meant that a flow is at its limit (or a flow on a line would be at its limit under contingency of another line—a so-called security constraint).

Whether there are binding transmission constraints or not, a principles-based assessment of market power instead focuses on the competitive situation faced by particular market participants at buses where they have economic activity, given the prevailing market conditions. The transmission constraints will affect these market conditions to determine the mark-up at each bus above competitive conditions.

2.1.5 Capital Formation

It might be argued that when transmission constraints bind, the limited options available to the ISO may mean that there is little ability to deviate from efficient operation, given available generation. (The discussion in Section 2.5.8 will provide a slightly different view.) Consequently, the main goal of market power mitigation in this case is to avoid excessive wealth transfers. Unfortunately, in such extreme circumstances, and analogously to the case of tight overall supply, it can be difficult to assess the competitive price and the resulting prices may be set far from competitive levels. This has great significance for long-term efficiency of markets, since investment in peaking generation, located in transmission-import constrained areas, is likely to be significantly affected by prices set when supply is tight and transmission constraints bind. Prices that are distorted away from competitive levels will result in poor signals for generation investment.

Poor signals for capital formation are likely to result in new generators being built in locations that are sub-optimal for managing transmission congestion, resulting in less efficient deployment of generation and the need for additional transmission. To summarize, market power, price formation, efficient operations, and efficient capital investment are all bound together.

2.1.6 Economic Significance of Inefficiency

As discussed in the last section, market power can affect capital formation, with long-term implications for efficiency of the electricity industry [1]. For example, withholding of supply may result in additional generation being built that would not have been necessary in a competitive industry. In this case, the inefficiency is due to both increased capital expenditures and deviations of operations from least cost. However, it is difficult to estimate the cost of such inefficiency because it involves both increased capital expenditure and inefficient production over the lifetime of the generation stock.

Even in the short-term operational domain without considering the effect on capital formation, empirical estimates of inefficiencies require detailed information about operating costs. Seminal studies on the exercise of market power in electricity markets include References [1, 3, 4]. For example, Reference [3] finds a price mark-up over competitive levels of around 20% in England and Wales in the late 1990s, while Reference [1] find levels of mark-up even larger in California in the period from 1998 to 2000 [1, Figure 3], and inefficiencies amounting to approximately 10% of operating costs in late 2000 [1, Table 4]. The inefficiency was a relatively smaller fraction of the total wholesale payments.

A recent example of an inefficiency estimate appears in Reference [5]. Similar to Reference [1], they also find production inefficiencies of approximately 10% of operating costs in Alberta [5, Section 4.1], although the inefficiency is again only a few percent as a fraction of wholesale payments. Nevertheless, such inefficiencies at this level are significant, being as large as the gains over many years of technological improvements in heat rates of generators.

2.1.7 Focus of Market Power Analysis and Mitigation

Efficiency of and market power in restructured electricity markets have both a short-term and a long-term character, with important interactions between the short-term operational domain, that is, the real-time market, and longer-term domains, including the day-ahead market and all other forward markets and trading opportunities in advance of the day-ahead market, and the installation and retirement of capital. An important observation is that prices in all the longer-term domains are influenced by expectations about the real-time market. Prices in the various markets are linked through forward contracting and (in the case of day-ahead and real-time markets) through "virtual bids" that allow for a sale into the day-ahead market and purchase of an equal quantity from the real-time market, or vice versa, at the respective prices in those markets. The implication is that by mitigating market power in the real-time market, market power will also be mitigated in the longer-term domains. For example, consider the withholding of supply (or of demand) from the day-ahead market when that supply (or demand) is anticipated to be present in real-time. Virtual bids will tend to compensate for the withholding in the forward market. This means that mitigation is typically best applied in the real-time market since it will have the most systematic effect throughout all markets, both real-time and forward, including markets for capital formation, through the influence of real-time prices on forward prices. (The interaction of forward contracting and real-time markets will be further discussed in Section 2.4.3.)

2.1.8 Summary of Remainder of Chapter

The rest of this paper will discuss market power and efficiency in detail. Price formation in electricity markets is described in Section 2.2. Price and offer caps are described in Section 2.3. The ability and incentive to exercise market power is described in Section 2.4. Market power mitigation approaches are described in Sections 2.5 and 2.6 concludes.

2.2 PRICE FORMATION IN ELECTRICITY MARKETS

2.2.1 Market-Clearing Prices

To discuss market power in detail, we must first define the formation of prices in restructured electricity markets. All North American day-ahead and real-time markets use so-called market-clearing prices. Electricity markets in North America and elsewhere require that most market participants with generation assets submit an offer to the Independent System Operator (ISO) for each of their generation assets; that is, a schedule of minimum accepted prices versus quantity is specified for each generator. Typically, there will also be various constraints on operation that are also specified by parameters as part of the offer, including ramp rate constraints, but we will mostly ignore these issues except for the maximum capacity of each generator.

The ISO dispatches offers by choosing to use lower priced offers in preference to higher priced offers. The algorithm is very similar to economic dispatch as practiced in vertically integrated utilities, and so it is called "offer-based economic dispatch," or "offer-based security-constrained economic dispatch" when consideration of transmission constraints is being emphasized [6].

As discussed in Reference [7], "putting aside demand bids and transmission constraints for simplicity," so that demand is fixed in a particular dispatch interval and transmission constraints are not binding, then "the market-clearing price is the offer price of the highest accepted offer in the market." As further discussed in Reference [7], an important reason for using market-clearing price is due to the incentives it provides for efficient dispatch and optimal investment. In particular, in a competitive market with market-clearing prices, meaning a market where generators cannot affect the price as determined by the offer price of the highest accepted offer, generators will maximize their profits by offering energy at prices closely reflecting their "marginal costs." The offer-based security-constrained economic dispatch algorithm performed by the ISO then results in operationally efficient dispatch, so that demand is served by resources having the overall lowest production costs.

In a competitive market, those generators with offer prices below the market-clearing price are paid more than their offer. The mark-up above competitive prices is zero under these competitive conditions. That is, the difference between the market-clearing price and an offer price does not constitute, in itself, market power.

The foregoing discussion considered the case where demand is fixed; however, the situation can be generalized to the case of demand bids that can allow for "dispatch" of demand, effectively meaning that some demand may not be served if the prices, or "willingness-to-pay" associated with its demand bids are insufficient. In particular, if generation offers and demand bids are considered together, then it can be the case that supply does not meet all of the "desired" bid demand.

If competitive demand bids set the market-clearing prices under those conditions when not all desired demand is served, then market prices will provide incentives for levels of investment that will maximize overall benefits of consumption minus the capital and operating costs of production, at least in the circumstances were we can ignore uncertainties and lumpiness of capital formation.

Uncertainties can be accommodated in this framework. In particular, if there are sufficiently liquid forward markets and certain other assumptions are made, then optimal investment will take place in expectation going forward. The various assumptions underpinning these observations may not be exactly satisfied in practice. Nevertheless, these assumptions provide a useful idealization that may be well-enough satisfied in practice that market participants will be guided toward near-optimal investment decisions. Optimal investment means that the right generation technologies are built in the right places at the right times. Efficient capital formation is an important reason justifying a desire for a competitive market, as supported by the single market-clearing price.

It is important to emphasize, however, that these favorable conditions include that prices are occasionally set by demand bids, or by a proxy to the demand willingness-to-pay. That is, occasional situations where not all desired demand is met is fundamental to the efficient capital formation argument. As will be further discussed below, this requirement has been problematic in restructured electricity markets.

2.2.2 Pay-as-Bid Prices

Despite the compelling reasons for using the single market-clearing price for electricity, alternative pricing rules are sometimes proposed. One such alternative proposal is "pay-as-bid," where each accepted offer is paid its offer price. As discussed in detail in Reference [7], "there is no empirical or experimental evidence that pay-as-bid or other alternatives would reduce prices significantly compared to a single market-clearing price design. In fact, some evidence suggests that pay-as-bid would increase prices compared to explicitly setting the single market-clearing price." Moreover, pay-as-bid has some significant drawbacks, including the potential for poor dispatch decisions in the practical case that market participants have only imperfect knowledge about other market participants. For these reasons, this chapter will only analyze the case of market-clearing prices.

2.2.3 Transmission Constraints

All North American markets represent transmission security constraints into the market-clearing conditions. When transmission constraints bind, market-clearing prices differ by bus. Under these circumstances, competition can be limited. Nevertheless, the ideal of competitive locational, marginal, or nodal prices provides similar favorable incentives for the formation of capital and operation of the system. Where we discuss transmission constraints, we will also explicitly consider nodal prices. Nodal prices are the generalization of market-clearing prices to the case of binding transmission constraints. Nodal prices are indeed market-clearing, reflecting the fact that market-clearing prices vary bus by bus since the prevailing market conditions vary from bus to bus.

2.2.4 Summary

To summarize, market-clearing prices have significant benefits in electricity markets when competitive conditions prevail. While pay-as-bid and other mechanisms are sometimes proposed to mitigate market power, this chapter will assume that market-clearing prices are used. That is, we will explicitly consider the market-clearing prices in analyzing market power and economic efficiency, including their locational character.

2.3 PRICE AND OFFER CAPS

As alluded to in Section 2.1, market participants will typically have most ability to influence prices when prices would, even under competitive conditions, be high. In the extreme of a pivotal supplier facing fixed demand, it could presumably ask for any price and receive it in the absence of market power mitigation. A natural approach to this issue is to limit the ability to further increase prices above competitive levels. An offer cap is a maximum value of price that can be submitted in an offer, as specified in the rules of the market.

A subtlety in the context of offer caps is that, in the presence of transmission constraints, it is possible for the highest locational price to exceed the highest offer price. If this is of concern, then prices can be truncated to a level set by a price cap. That is, the offer cap limits the highest price specified in any offer, while a price cap limits the highest price that is actually paid in the market.

While setting price and offer caps can prevent excessive transfers, it may also limit prices to being below competitive levels if the caps are set too low. The setting of such caps is therefore inevitably a compromise between preventing too high prices under conditions where market power prevails, while allowing the prices to get high enough to remunerate efficient levels of investment. If caps are set too high, excessive transfers can result, while if they are too low, insufficient investment will occur, which will presumably exacerbate market power in the future if demand grows over time.

Because of the difficulty of setting levels of caps to accommodate both the function of preventing excessive transfers and also allowing for competitive prices, caps should generally be considered a fairly coarse mitigation tool. Price and offer caps provide a backstop against excessive transfers, but cannot typically be relied upon to appropriately mitigate market power or achieve competitive conditions. For example, caps in several markets are at levels that represent the notional dis-benefit of involuntary curtailment or blackout, the so-called "value of lost load" (VOLL). Presumably no one is prepared to pay more than VOLL for energy under any circumstances; however, many consumers might be willing to choose to not consume at lower prices than VOLL, given appropriate circumstances.

For the rest of the chapter, it will be assumed that offer and price caps are in place as a backstop. However, other explicit mechanisms must be in place to fully mitigate market power and achieve competitive outcomes.

2.4 ABILITY AND INCENTIVE TO EXERCISE MARKET POWER

As mentioned in Section 2.1, under tight supply conditions, and particularly when transmission constraints bind, conditions will typically not be competitive. That is, at least some market participants will be able to influence the price. To understand this more fully, we will consider the ability, incentives, and circumstances in which a market participant may raise its offers above competitive levels using a stylized model of the behavior of market participants. This section will provide qualitative insights into the circumstances that are associated with the ability and incentive to exercise market power.

To understand incentives and behavior, it is necessary to model the motivations of market participants. Although various short- and long-term issues may affect decision-making, a useful, but stylized, guide to actions is to consider what actions would maximize profits as tacitly assumed in Section 2.2, where it was observed that competitive market participants in a clearing-price market are motivated to offer at their marginal cost since this maximizes their profits.

In the context of real-time markets that are typically repeated on a five minute-by-five minute basis to follow changes in demand, the profit in this context is the short-term operational profit. The following analysis of profit maximization is standard, with notation and discussion borrowed from Reference [8]. The development will lead to an index of market power; however, several other indices appear in the literature, including the consideration of "residual supply" [9–12].

Consider a firm i that owns generators at buses $k \in K$, where $|K| = r$. We collect the production quantities $q_k, k \in K$, at all these generators into a vector $q \in \mathbb{R}^r$. There are also assumed to be other firms. All firms are required to make offers into the market. However, if we fix the offers of all firms but firm i to focus on the actions of firm i, then we can observe that if the offers associated with firm i's generation were varied then this would result in variation of the generation levels for firm i in the offer-based security-constrained economic dispatch as determined by the ISO. The varying dispatch levels will also be associated with varying price levels. We can therefore think of the implicit relationship between the resulting dispatch levels q and the resulting prices at the corresponding buses and conceptually solve for the function p_k that specifies the price at bus k as a function of the generation q of firm

i, assuming that offers of all other firms are held fixed. That is, we assume that a function $p_k : \mathbb{R}^r \to R$ can be found that evaluates the market-clearing price at bus k given that firm i submitted an offer such that the ISO then dispatched firm i to produce the quantities q. The function p_k is called the inverse residual demand function for firm i at bus k. We collect the inverse residual demand functions $p_k, k \in K$, together into a vector function $p : \mathbb{R}^r \to \mathbb{R}^r$ that reflects the prices at all buses where firm i owns generation.

Note that the inverse residual demand faced by a generator at bus k depends on the whole vector q. That is, actions by the firm at any one of its generators may result in a change in the price at bus k. Moreover, this effect may be due to the interaction of multiple binding transmission constraints, and generally speaking as more constraints become binding the sensitivity of price to generation increases in magnitude: a given change in generation at a bus results in a larger change in price.

In the simple case that $r = 1$, so that the firm only owns one generator (or only owns generators at one location) the inverse residual demand has the property that its slope, the inverse residual demand derivative, is non-positive. That is, increasing production by the generator will not result in increasing prices.

The generalization of the slope to the case $r > 1$ is called the inverse residual demand Jacobian; that is, the r by r matrix of partial derivatives of prices with respect to generation quantities. The diagonal entries of the inverse residual demand Jacobian, $\frac{\partial p_k}{\partial q_k}, k \in K$, reflect the sensitivity of price at a bus k to generation at bus k and are non-positive. The off-diagonal entries, $\frac{\partial p_k}{\partial q_m}, k \neq m$, for $k, m \in K$, which represent the sensitivity of price at bus k to generation at another bus m, may be either positive or negative. However, the inverse residual demand Jacobian has the property that it is a negative semi-definite matrix [13, 14].

The definition of market power in Section 2.1 is that the market participant can affect prices. Interpreting this definition in terms of the inverse residual demand Jacobian, a market participant has market power if the entries of the Jacobian are significantly different from zero since these entries measure the sensitivity of prices to the quantities q, and the quantities q are implicitly determined by the offers of firm i.

The measure of market power defined in Section 2.1 is the mark-up above competitive price. We will approximate this mark-up by applying the assumption that firms are profit maximizers and considering the condition for firm i to maximize its profit. Assume that the production cost functions of the firm are specified by the cost functions $c_k : R \to R, k \in K$. Ignoring forward contracts for now, the profit for the market participant is:

$$\forall q \in \mathbb{R}^r, \pi(q) = \sum_{k \in \mathbb{R}} q_k p_k(q) - c_k(q_k),$$

where the term $\sum_{k \in K} q_k p_k(q)$ is the total revenue of firm i. (The case with exogenously specified forward contracts is similar.) We first consider the case where the capacity constraints of the generators owned by the firm are not binding, then qualitatively describe the more general case of capacity constraints, and finally consider forward contracts.

2.4.1 Ignoring Generator Capacity Constraints

To simplify the discussion, we will assume that sufficient conditions for maximization of profit are satisfied, that p and $c_k, k \in R$, are differentiable, and that generator capacity constraints are not binding. In this case, we can find necessary conditions to maximize the profit by setting the partial derivatives of profit with respect to quantities equal to zero. Focusing on the partial derivative with respect to q_m for a particular $m \in K$, we obtain:

$$0 = \frac{\partial \pi}{\partial q_m}(q) = p_m(q) + \sum_{k \in K} q_k \frac{\partial p_k}{\partial q_m}(q) - c'_m(q_m),$$

where $c'_m = \frac{\partial c_m}{\partial q_m}$ is the marginal cost of the generator owned by the firm at bus m. Re-arranging the above equation, we obtain the price-cost mark-up at bus m under the hypothesis that the firm was maximizing its profits:

$$p_m(q) - c'_m(q_m) = -\sum_{k \in K} q_k \frac{\partial p_k}{\partial q_m}(q). \tag{2.1}$$

This is a generalization of (1) in Reference [15] to the case of firms with multiple generators. For the special case of $r = 1$, where firm i only owns generation at a single bus m, (2.1) specifies that the difference between price and firm marginal cost is $q_m \frac{\partial p_m}{\partial q_m}(q_m)$, which depends only on the generation q_m at bus m (and implicitly on the offers from the other firms). In the case of a firm with generators at multiple buses, the profit maximizing mark-up at generator m depends on productions at other buses q_k, as well as on q_m, and depends on the values of the cross partial derivatives $\frac{\partial p_k}{\partial q_m}$.

To aggregate all the estimated mark-ups into one index for firm i, Reference [8] evaluates the quantity-weighted estimated average mark-up of the firm, which results in a particularly convenient expression:

$$\left(-q^\dagger \frac{\partial p}{\partial q}(q)q\right) \Big/ \left(\sum_{k \in K} q_k\right), \tag{2.2}$$

where $\frac{\partial p}{\partial q}(q)$ is the inverse residual demand Jacobian evaluated at q, and superscript $\dagger$ means transpose.

Since cross-derivatives $\frac{\partial p_k}{\partial q_m}$ for $k \neq m$ can be positive, it may be the case that, at some buses, profit maximization corresponds to a "mark-down" rather than a mark-up. That is, for some buses m, the estimated mark-up in (2.1) may be negative. While this seems to be counter-intuitive, Hogan [16] and Cardell et al. [17] describe just such a case where a firm offers below marginal cost at bus e on the exporting side of a transmission constraint in order to congest the line and consequently be able to offer well above marginal cost at a bus m on the importing side. That is, the mark-up considered at any given bus does not give a full picture of the situation faced by a firm when transmission constraints bind. However, the index (2.2) combines the effect of mark-up at all buses. Since the Jacobian is negative semi-definite, the quantity-weighted mark-up is non-negative. Moreover, if generation costs are convex and capacity constraints are not binding, then the necessary conditions are sufficient.

To summarize, this estimate assumes that the firm is maximizing its profits and that the firm can evaluate the residual demand it faces. As discussed in [15], even given these assumptions, these must be viewed as only approximate estimates of mark-up over *competitive prices,* since competitive prices at each bus m may deviate from the marginal costs $c'_m(q_m)$ at the market-clearing conditions.

Nevertheless, the index provides some qualitative insights. The Jacobian entries in the inverse residual demand derivative are non-zero wherever firm i can affect prices. If the entries are large in magnitude, then the mark-up will be large, reflecting profit maximizing exercise of market power and its withholding of its generation capacity. Typically, the marginal cost of generation at bus k owned by firm i at its production level q_k will be below competitive levels whenever generation is withheld, but the market prices will be above competitive levels, so some other firms will typically be producing more than under competitive conditions.

In summary, production will be inefficient because of inefficiently low generation by firm i (and by other firms with market power) and there will be inefficiently high generation by some or all of the firms that do not have market power. Overall operating costs could be reduced by transferring production from higher marginal cost generators to lower marginal cost generators, as in the standard notion of bringing dispatch toward conditions of economic dispatch by backing of production at high marginal cost generators and increasing production at low marginal cost generators.

2.4.2   Considering Generator Capacity Constraints

The case of generators at their maximum capacity is somewhat more complicated and is described in detail in Reference [8]. However, similar results prevail: inefficiently low production by firms with market power, and inefficiently high production by some other market participants. The added complexity is due to the issue that a fully-dispatched generator is not withholding, but the price at its bus may be above competitive due to withholding at other generators owned by the same market participant.

2.4.3   Forward Contracts

Forward contracts affect the situation by changing the exposure to the market prices. This can either reduce or amplify the exercise of market power depending on the forward contract position. In particular, if firm i sells forward a quantity $q_k^f \geq 0$ of energy from generator k at an agreed price p_k^f then the exposure to the real-time price is reduced because the revenue for generator k is changed from $q_k p_k(q)$ to $(q_k - q_k^f) p_k(q) + q_k^f p_k^f$, reflecting the remuneration of the quantity q_k^f at the forward price p_k^f, and of the quantity $q_k - q_k^f$ at the real-time price $p_k(q)$.

We can still consider the profit maximizing conditions given the modified revenue in the real-time market. However, profit maximization now involves a smaller mark-up than in the case of no forward selling. Conversely, if the generating firm buys energy forward then its exposure to real-time prices has increased, and its profit maximizing mark-up will be higher than in the absence of forward contracting.

Given the above analysis, it might seem that firms with generation would be inclined to not sell energy forward so as to preserve their market power in the real-time market. Potentially, they might be expected to even buy energy forward in order to amplify their market power. However, real-time prices are very volatile, and a generator selling forward mitigates its exposure to this risk, whereas buying forward would increase exposure. Moreover, longer-term forward selling can provide collateral for new generation construction. That is, for several reasons, generating firms may still be inclined to sell forward even if selling forward mitigates their market power.

There are even stronger reasons to expect generators to sell energy in forward markets due to an important result of Allaz and Vila [18]. In particular, their work shows that when market participants can participate in forward and real-time markets, the resulting forward contract level will tend to reduce the market power compared to a situation with only a real-time market because of the interaction between the markets.

Nevertheless, even if market power in the real-time market is partly ameliorated by interaction with forward markets, it will generally not be completely eliminated by this interaction. Because prices in forward markets are generally based on expectations about real-time prices unless there are barriers to arbitrage between these markets, market power in real-time markets will result in prices in forward markets that are also distorted. As mentioned above, longer-term forward markets have important roles in capital formation, since they can provide collateral for financing. Consequently, distorted forward prices due to expectations of market power in real-time markets will contribute to poor capital investment decisions and inefficient investment. This confirms that mitigation of market power should usually focus on real-time markets since a competitive real-time market will also "discipline" the forward markets, including the markets that influence capital formation, toward competitiveness.

2.5 MARKET POWER MITIGATION APPROACHES

This section will describe several market power mitigation approaches. They will be evaluated from the perspective of their intention and ability to result in competitive prices and conditions. Section 2.5.1 reiterates the backstop function of price and offer caps. Section 2.5.2 describes conduct and impact tests, while Sections 2.5.3 and 2.5.4 describe structural tests and the "Texas two-step," respectively. Section 2.5.5 describes limitations on changing of parameters and Section 2.5.6 describes transmission right ownership rules. All of these approaches are supply-side mitigation approaches and, as will be described, can help to provide competitive prices absent scarcity. They do not generally result in competitive levels of prices under scarcity because they do not represent demand willingness-to-pay.

Section 2.5.7 describes demand response, while Section 2.5.8 describes "operating reserve demand curves" and other administrative proxies for demand response. Demand response and proxies to it have an important role in reducing market power. These demand-side approaches also handle scarcity conditions since, by definition, they result in less than the desired demand being served, or in some cases result in less ancillary services being provided.

2.5.1 Price and Offer Caps

As discussed in Section 2.3, price and offer caps are fairly coarse market power mitigation methods, with limited ability to result in competitive price conditions. Nevertheless, they provide a useful "backstop" function that can guard against excessive transfers if the other approaches are not fully effective.

2.5.2 Conduct and Impact Tests

Conduct and impact tests form the basis of mitigation in several markets, including the "Automated Mitigation Procedure" of the New York ISO [19]. As the name implies, these tests consist of a conduct assessment and an impact assessment.

The "conduct" assessment first investigates whether offers from a market participant are at significantly higher prices than a "reference" offer for the participant. The reference offer is based, for example, on historical offer behavior by the market participant, using the lowest previous price at the offer location, or on fuel costs and an assumed input-output characteristic of each generator [20]. If competitive pressures apply during the historical period, then it can be assumed that the offers will reflect marginal operating costs. If competitive pressures apply during only some of the historical record, then the reference offer should be based on the circumstances when the offers were competitive. The conduct test compares the actual to the reference offer and if the actual offer is sufficiently higher than the reference offer then the test proceeds to the impact assessment.

The "impact" assessment determines whether the higher offer by the market participant made a difference to the market outcome compared to a counterfactual case with the reference offer. The market clearing algorithm is re-run with all the offers of other market participants kept at their actual levels, but the offer of the market participant in question is changed to be equal to the reference offer. If the two market outcomes are materially different, then the offer is "reset" to the reference offer.

As mentioned above, the reference offer could be based on historical behavior or on fuel costs and an assumed input–output characteristic of each generator. Market power mitigation based on historical data would be successful if the historical offer data for all generators included times when they were exposed to competitive conditions that resulted in offers at close to marginal costs. While this may be true for most generators, it is unlikely to be fully realized for peaking units and generators in import-constrained regions that are only dispatched during particular high load conditions. Unfortunately, this is precisely the condition when supply is close to scarce, either throughout the system or locally, and so conditions are almost certainly not competitive. If the historical conduct of a peaking generator has always been to offer at a high price, then evaluating the reference price using historical behavior will not reveal the marginal costs and the conduct test may never trigger an evaluation of the impact.

Since generator input–output curves are fairly well understood for most technologies, the use of fuel costs and input–output curves is possibly more reliable than the use of historical offer data to evaluate reference offers. Modern combined-cycle units, for example, have standardized combustion turbines and the overall heat rate can be reliably estimated.

In summary, the conduct and impact test is least reliable for those generators that are called upon infrequently under scarcity or close-to-scarcity conditions. These are precisely the conditions, however, when mitigation is most likely to be needed. In practice, the markets that have adopted conduct and impact tests also have so-called "capacity markets" that effectively avoid scarcity under most conditions by providing for planned investment to cover foreseeable demand levels. Under conditions where there is sufficient generation capacity to meet desired demand, the conduct and impact approach can reasonably mitigate prices to reflect marginal costs.

2.5.3 Structural Tests

In the structural approach, offers are mitigated when particular market concentration conditions are satisfied in the context of locational market power [20]. For example, PJM has a three pivotal supplier test, which applies mitigation when the capacity of the three largest suppliers affecting a particular transmission constraint is necessary to keep the flow within limits, evaluating that constraint as "non-competitive." Offers are then mitigated to the reference offer, analogously to the case of the conduct and impact test described in Section 2.5.2.

CAISO has a similar test. Similarly, the "Texas two-step" to be discussed in the next section attempts to evaluate which constraints are "competitive" and which are "non-competitive." As discussed in Section 2.1, the paradigm of individually identifying constraints as "competitive" or "non-competitive" ignores the interaction of multiple binding constraints.

This approach can certainly limit prices; however, a literal interpretation of the test is tantamount to assuming that three firms would collude to set prices, a serious violation of United States antitrust law. This is presumably aimed as a proxy to detecting conditions that might be conducive to the exercise of unilateral market power. In the absence of scarcity, it will presumably limit offers to being competitive, but is unlikely to result in competitive prices in the presence of scarcity, absent demand bids.

2.5.4 Texas Two-Step

The Texas two-step is utilized in the Electric Reliability Council of Texas (ERCOT) as an approach to mitigating locational market power in the real-time market, and has some of the flavor of the structural test described in Section 2.5.3. (Other market use somewhat different mitigation procedures for locational market power.) Constraints are divided into "competitive" constraints and "non-competitive" constraints on the basis of constraint-by-constraint evaluation. As with the structural tests described in Section 2.5.3, this paradigm ignores the interaction of multiple binding constraints. In practice, the constraints that are evaluated to be competitive in ERCOT roughly correspond to the inter-zonal constraints in the previous zonal market, while all other constraints are evaluated to be non-competitive.

In the first stage of the Texas two-step, only the competitive transmission constraints are represented in the market-clearing algorithm. This results in reference prices. In the second step, offers are reset if they exceed the higher of:

- the reference prices, and
- unit-specific offer caps.

The offers are then reset to this maximum of the reference prices and the unit-specific offer caps for the second stage. In the second stage, all transmission constraints are represented in the market-clearing algorithm, but the high offers have been reset based on the first stage results. The rationale is that if there is zonal-wide scarcity then the reference prices in the first stage will be high, allowing for high prices in the second stage. On the other hand, if there is not zonal-wide scarcity, then offers will be mitigated for the second stage.

As with structural tests, this approach will certainly limit prices; however, in circumstances of intra-zonal, that is, local scarcity it will typically result in prices that are below competitive. Currently, the ERCOT market only partially supports active demand-side bids in the real-time market, so under scarcity conditions prices are not typically set by the willingness-to-pay of demand-side resources. Real-time bidding is implemented in other markets and enhancements of real-time bidding is under discussion in ERCOT and would allow for closer to competitive prices in the case of scarcity. Again, the Texas two-step alone does not provide competitive prices in the testing case of scarcity associated with binding intra-zonal transmission constraints, but works reasonably well in the absence of scarcity and in the case of zonal scarcity to mitigate prices to marginal costs.

2.5.5   Limitations on Changing of Parameters

The discussion of market clearing has tacitly assumed that an offer is in one-to-one correspondence with the demand conditions. That is, it has implicitly assumed that offers can be updated for each market-clearing interval. In practice, in real-time markets, there are typically either four 15-minute or twelve 5-minute clearing intervals, with most North American markets now utilizing twelve 5-minute clearing intervals per hour. Offers into the real-time market are fixed for each hour. That is, a single offer applies to all 12 clearing intervals.

As discussed in Reference [21], when offers are required to stay fixed over multiple clearing intervals during which demand levels change significantly, this will tend to mitigate market power. The basic reason is that the single offer must compromise between low and high demand conditions. The wider the range of demand, the more effectively is market power mitigated. While there is some variation of demand over an hour, the daily variation is much larger. Some markets, such as PJM, restrict most offers in the day-ahead market to be fixed for all hours, while other markets such as ERCOT have looser restrictions.

However, the insight that limits on changing of parameters can mitigate market power has been utilized in a so-called "voluntary mitigation plan" agreed between NRG and the Public Utility Commission of Texas for NRG's participation in the ERCOT market [22]. In this agreement, there was some allowance for offers that would be well above the typical marginal cost, but there was a requirement for these offers (and other parameters reflecting generator capability) be held fixed for all 24 hours. Over such an extended period, demand varies significantly in ERCOT, and the effect is to mitigate the market power. With increased variability due to renewables, intra-hour and daily variability of net demand will also increase.

While this approach can help to mitigate market power, if the limitations on changing parameters are too strict, they can interfere with the ability to reflect valid changes, thereby degrading market efficiency and affecting system reliability. For example, suppose that a firm is limited from changing the specified ramp rate limits of its generators. Furthermore, suppose that for a particular generator, the firm specified a representative limit for all 24 hours of the day that was nevertheless too optimistic under particular conditions. In this case, the generator might be dispatched to ramp at its stated ramp rate limit, but be physically incapable of responding to the deployment. In this case, the price would be depressed because the next higher cost resource that could actually respond should have been dispatched, while reliability would be affected because the resource that was dispatched cannot actually respond. Another example occurs if fuel costs vary within the day; however, such a situation can be handled by allowing for separate specification of fuel offer costs and efficiency of conversion of fuel into energy through an input–output curve, as described in the context of conduct and impact tests.

2.5.6 Financial Transmission Right Ownership Rules

Financial transmission rights (FTRs) are issued by ISOs to provide hedging of nodal price differences when transmission constraints bind. Nodal price differences are the short-term costs to transmit power between buses. Just as nodal prices are volatile, nodal price differences are also volatile, and there is a desire to hedge this volatility.

FTRs are effectively pairs of forward contracts involving an offer to sell and bid to buy. Just as individual forward contracts can amplify or mitigate market power, FTRs can also do so depending on whether they reduce or increase the exposure to real-time prices. One approach to market power mitigation then is to limit the ability of market participants to hold FTRs that amplify their market power [23]. In practice, markets have limitations on holding of FTRs that are independent of the specific effect on market power.

2.5.7 Demand Response

Demand response can be divided into two types: passive and active. In passive demand response, a load (or the load-serving entity that represents the load) anticipates or experiences a high price and decides to "self-curtail" its consumption. This type of response is analogous to finding prices too high in a supermarket and deciding not to buy. Such behavior can help to mitigate market power, by reducing the quantity consumed at the high price: profitability of offering at a high price is reduced. Another way to think of such demand response is that it reduces the magnitudes of the entries in the inverse residual demand Jacobian.

A drawback of passive demand response is that the resulting prices are still determined by supply offers. Generally speaking, such offers must be significantly above the typical marginal cost of generation if they are to induce passive demand response. That is, passive demand response mitigates market power, but also depends on there being some market participants with enough market power to actually result in the high prices that will induce self-curtailment. Nevertheless, if there are enough offers at varying prices, and enough passive demand response, then the combination of the resulting generation, the actually served demand, and resulting price can approximate competitive conditions.

However, there are two further problems with passive demand response. First, if the high offer prices that induce passive demand response are sparsely separated, then the passive demand response may be lumpy, which can potentially complicate supply–demand balance in the system. (An analogous issue arises with large "all-or-nothing" loads; that is, loads that are either on or fully off, functioning as ancillary services.) In Section 2.5.8, an additional aspect of demand response will be described that can "fill-in" the range of prices more effectively and reduce the severity of this issue, while at the same time enhancing the market power mitigation effect.

Second, it may be the case that passive demand response changes over time. For example, high prices may induce reduced use of air conditioners only temporarily. When the load "rebounds," it may become less elastic and thereby allow for the exercise of market power.

In active demand response, load (or representatives of load) explicitly bids the willingness-to-pay into the market. The bids are represented in the optimal dispatch problem, and the bid load is dispatched analogously to generation, although sometimes with specific allowances such as "all-or-nothing" bids. In active demand response, if only part of a desired demand is fulfilled, then the willingness-to-pay of the corresponding bid will set the market-clearing price. Active demand response requires that demand-side consumption can be controlled so that it can be "dispatched down." This is often easiest for industrial processes, or large commercial consumers. However, there is also potential for aggregation of loads such as air conditioners, water heaters, and plug-in electric vehicles to be dispatched.

As with passive demand response, the willingness-to-pay of such active demand bids is typically much larger than the typical marginal cost of generation resources, so that, under competitive conditions, generation resources will almost always be fully deployed before demand would be dispatched down. Several ISO markets allow for demand side bidding of willingness-to-pay and it has a very important role in mitigating market power and achieving efficient dispatch. In contrast to the supply side mitigation discussed in Sections 2.5.1–2.5.6, active demand response can mitigate market power even when generation is fully or close-to-fully dispatched. In particular, the demand response will reduce the magnitude of the terms in the inverse residual demand Jacobian (in the extreme of otherwise pivotal suppliers, active demand bidding can prevent the terms in the inverse residual demand Jacobian from becoming infinite), thereby reducing the incentive to increase offer prices.

2.5.8 Proxies to Demand Response

If there is a significant amount of passive demand response, then the ISO may find that a short-term forecast based only on non-price explanatory variables is in error. For example, the short-term forecast for the next dispatch interval may indicate an increase in demand compared to the current interval, requiring that an additional high-priced offer be dispatched and therefore set the market-clearing price. However, if passive price responsive load can anticipate or is informed of this price rise, then it may choose to not consume, and the forecasted increase may not materialize. This can result in over-generation and the consequent need for regulation ancillary services (AS) to compensate for the supply–demand imbalance.

A natural approach to improving the short-term forecast is to incorporate an estimate of the level of passive response in the clearing algorithm. That is, a proxy to the demand response could be included in the ISO dispatch algorithm, which has the effect of making the market power mitigating effect of the demand response more explicit in the market-clearing conditions, as compared to being implicit in passive demand response behavior over successive clearing intervals.[3]

Considering the cost and value of AS allows for an additional dimension of representation of a proxy to demand response, with additional benefits in mitigating market power. AS such as reserves and up regulation require generation capacity. This capacity is not available for production of energy (unless the reserves are actually deployed.) The total generation capacity is divided amongst the provision of energy and of such AS. In initial implementations of electricity markets, AS requirements were set as fixed quantities. However, this does not recognize the fact that as supply becomes tight, it may be preferable to deplete reserves rather than curtail demand. That is, it is typically more efficient to deplete reserves than to immediately curtail demand. The "cost" of depleting the reserves depends on an assessment of the likelihood that operating with lower levels of reserves will result in involuntary curtailment due to a forced outage or other event, together with an assessment of the dis-benefit of

---

[3] Some customers in ERCOT have a tariff for transmission service that is based on their contribution to four peak demand events in the Summer. Large consumers anticipate these so-called "4CP" events and adjust their demand accordingly, demonstrating that such passive demand response is possible. Moreover, ERCOT anticipates this specific form of "demand response" in its short-term forecast.

the resulting involuntary curtailment, the VOLL, if the forced outage were to actually occur.

This insight is implemented in the so-called "operating reserves demand curve" (ORDC). Instead of having fixed requirements for AS, there is a willingness-to-pay specified for various levels of AS. The bid prices for AS are set, in principle, equal to the product of VOLL and the probability that load will be involuntarily curtailed at that level of procured AS. This probability is called the "loss of load probability" (LOLP). That is, the bid price represents the marginal expected value of the lost load.

At a sufficiently high level of AS, the LOLP is negligible and the full "desired" amount of AS would be procured, as in the initial implementations of AS markets. However, if supply becomes tight, a lower level of AS will be procured, and the co-optimization of provision of generation capacity for both energy and AS will then result in prices of both energy and AS increasing compared to the case of plentiful supply.

Since the LOLP typically increases fairly smoothly with declining quantity of AS, the resulting prices will smoothly cover a range from essentially zero to close to VOLL. That is, the ORDC approach can provide for a "filling in" of the range of demand response prices, which can facilitate passive demand response. In particular, in a hypothetical situation where supply became tighter throughout the day, prices would increase relatively smoothly and predictably over the day, providing opportunity for consumers exposed to wholesale prices (or their retailers) to respond incrementally with each successive increase in price as implied by the ORDC.

Even if the demand side does not respond significantly to the higher prices, analogously to demand response discussed in Section 2.5.7, the ORDC has the effect of decreasing the magnitude of terms in the residual demand Jacobian of market participants. That is, it mitigates market power even if demand does not respond directly to it. In summary, the combination of the ORDC and any passive demand response can significantly increase the overall price responsiveness and help to mitigate market power.

A related observation applies to the enforcement of transmission security constraints. Enforcement of such constraints in the market-clearing mechanism is aimed at avoiding system operations risk due to transmission contingencies. While these limitations are sometimes described as "hard" constraints that must be enforced, they can typically be relaxed somewhat. Thermal capacities, in particular, represent underlying assumptions about the duration of a flow on a line until automatic or operator action can be relied upon to reduce the flow to lower, sustainable levels. Higher limits on flows can be accommodated, but require faster action. Implicitly, such requirements for faster action come with a greater risk that operator actions will require involuntary curtailment of load.

In practice, such thermal constraints are modeled in dispatch algorithms as "soft" constraints, which can be violated at some cost. Implicitly, the relaxation of the constraints means that the constraints are somewhat elastic. Consequently, appropriate choice of the representation of these constraints can effectively provide further elasticity of demand to the market even under transmission-constrained conditions. This also serves to mitigate market power.

2.6 CONCLUSION

This chapter has discussed market power and its relationship to short- and long-term efficiency, emphasizing the importance of mitigating market power in real-time markets because of its influence on all forward markets. Although avoiding excessive transfers is often articulated as a goal of market power mitigation, this chapter has emphasized the nexus between mitigating market power and improving efficiency, both in the short-term operational context and also in longer-term contexts, particularly including capital formation: efficient markets will tend to build the right amount of generation capacity in the right locations at the right time.

The basic measure of market power adopted in this chapter is the mark-up above competitive levels. Analysis of the incentives to exercise market power reveals how exercise of market power is also generally associated with inefficient production. It will typically also be associated with inefficient capital formation. In situations where market power is significant and long-lasting, generation may be overbuilt in some contexts, while over-mitigation of market power may result in insufficient generation being built.

Market power mitigation can be aimed at influencing the supply side, or increasing the discipline provided by the demand side. Supply side mitigation approaches are generally able to reduce prices to approximate competitive levels, at least in the absence of scarcity. However, supply side techniques are not generally useful in the context of scarcity, where there is insufficient generation to meet all the desired demand. Although this situation is avoided in many North American market constructs through the operation of capacity markets, it is ultimately the case that greater demand response, and proxies to demand response, help to mitigate market power and ensure market efficiency in both the short- and long-run.

ACKNOWLEDGMENTS

The author would like to thank Mr. Daniel Jones, former Independent Market Monitor of ERCOT, for discussions and comments on this paper.

REFERENCES

[1] S. Borenstein, J. B. Bushnell, and F. A. Wolak, "Measuring market inefficiencies in California's restructured wholesale electricity market," *The American Economic Review*, vol. 92, no. 5, pp. 1376–1405, December 2002.

[2] M. H. Rothkopf, "Daily repetition: a neglected factor in the analysis of electricity auctions," *The Electricity Journal*, vol. 12, no. 3, pp. 60–70, April 1999.

[3] C. Wolfram, "Measuring duopoly power in the British electricity spot market," *The American Economic Review*, vol. 89, no. 4, pp. 805–826, September 1999.

[4] P. Joskow and E. Kahn, "A quantitative analysis of pricing behavior in California's wholesale electricity market during Summer 2000," *The Energy Journal*, vol. 23, no. 4, pp. 1–35, 2002.

[5] D. P. Brown and D. E. H. Olmstead, "Measuring market power and the efficiency of Alberta's restructured electricity market: an energy-only market design," 2015. Available at

https://econ.ucalgary.ca/event/2015-10-23/measuring-market-power-and-efficiency-albertas-restructured-electricity-market

[6] M. C. Caramanis, R. E. Bohn, and F. C. Schweppe, "Optimal spot pricing: practice and theory," *IEEE Transactions on Power Apparatus and Systems*, vol. PAS-101, no. 9, pp. 3234–3245, September 1982.

[7] R. Baldick, "Single clearing price in electricity markets,", February 2009. Available at http://www.competecoalition.com/files/Baldick%20study.pdf

[8] Y.-Y. Lee, R. Baldick, and J. Hur, "Firm-based measurements of market power in transmission-constrained electricity markets," *IEEE Transactions on Power Systems*, vol. 26, no. 4, pp. 1962–1970, November 2011.

[9] California ISO, "Residual supply metrics: methodology and preliminary 2009 results," 2010. Available at http://www.caiso.com/2725/2725e3899550.pdf. Accessed July 2, 2010.

[10] A. Sheffrin and J. Chen, "Predicting market power in wholesale electricity markets," Proceedings, 15th Annual Western Conference of the Advanced Workshop in Regulation and Competition, South Lake Tahoe, CA, 2002.

[11] A. Sheffrin, J. Chen, and B. Hobbs, "Watching watts to prevent abuse of power," *IEEE Power and Energy Magazine*, vol. 2, no. 4, pp. 58–65, July/August 2004.

[12] S. Bose, C. Wu, Y. Xu, A. Wierman, and H. Mohsenian-Rad, "A unifying market power measure for deregulated transmission-constrained electricity markets," *IEEE Transactions on Power Systems*, vol. 30, no. 5, pp. 2338–2348, September 2015.

[13] J. Barquín, "Symmetry properties of conjectural price responses," Proceedings of the IEEE Power and Energy Society General Meeting, Pittsburgh, Pennsylvania, July 2008.

[14] L. Xu and R. Baldick, "Transmission-constrained inverse residual demand Jacobian matrix in electricity markets," *IEEE Transactons on Power Systems*, vol. 26, no. 4, pp. 2311–2318, November 2011.

[15] Y.-Y. Lee, J. Hur, R. Baldick, and S. Pineda, "New indices of market power in transmission-constrained electricity markets," *IEEE Transactions on Power Systems*, vol. 26, no. 2, pp. 681–689, May 2011.

[16] W. W. Hogan, "A market power model with strategic interaction in electricity markets," *The Energy Journal*, vol. 18, no. 4, pp. 107–141, 1997.

[17] J. B. Cardell, C. C. Hitt, and W. W. Hogan, "Market power and strategic interaction in electricity networks," *Resource and Energy Economics*, vol. 19, no. 1–2, pp. 109–137, March 1997.

[18] B. Allaz and J. L. Vila, "Cournot competition, forward markets, and efficiency," *Journal of Economic Theory*, vol. 59, no. 1, pp. 1–16, 1993.

[19] P. Adib and D. Hurlburt, "Market power and market monitoring," in *Competitive Electricity Markets: Design, Implementation, Performance*, F. P. Sioshansi, Ed., Chapter 7, pp. 267–296. Elsevier Science Ltd., 2008.

[20] Federal Energy Regulatory Commission, "Price formation in organized wholesale electricity markets," Docket No. AD14-14-000, October 2014. Available at www.ferc.gov

[21] R. Baldick and W. Hogan, "Stability of supply function equilibrium: implications for daily versus hourly bids in a poolco market," *Journal of Regulatory Economics*, vol. 30, no. 2, pp. 119–139, 2006.

[22] Public Utility Commission of Texas, "Settlement agreement and voluntary mitigation plan pursuant to PURA s(1)5.023(f) and P.U.C Substantive Rule 25.504(e)," Docket Number 40488, 2012. Available at http://interchange.puc.texas.gov

[23] M. Joung and R. Baldick, "The competitive effects of ownership of financial transmission rights in a deregulated electricity industry," *The Energy Journal*, vol. 29, no. 2, pp. 165–184, 2007.

PART II

UNDER SMART GRID ERA

CHAPTER 3

MASS MARKET DEMAND RESPONSE MANAGEMENT FOR THE SMART GRID

Alex D. Papalexopoulos

3.1 OVERVIEW

The increasing complexity of the electricity industry requires the need to better understand and manage diverse resources like renewable energy sources (RES), such as solar and wind energy, renewable hydroelectricity, demand response (DR), energy storage resources (ESR) and distributed energy resources (DERs). The looming development of storage and demand-side generation technologies, including plug-in hybrid vehicles, is already affecting plans for future power management. Using technology to improve our ability to analyze the grid and manage power generation, transmission and delivery in the most efficient manner is imperative. End-users must be empowered as well—armed with timely information such as real-time time-of-use pricing—to guide their energy use decisions. The "smart grid" is all about better information and communication and the ability to make wise decisions about energy usage at both micro and macro levels.

Smart meters, big data, analytics, cloud-based technology and the Internet of Things (IoTs) are all making contributions to a smarter energy and grid system, which are likely to revolutionize the energy sector. The combined impact is expected to rival that already seen in telecoms or banking, where much of the technology has its origins. Development is also being driven by regulatory changes to allow customer choice in countries with competitive markets, along with the need to accommodate more intermittent RES, major offtake points, and Distributed Generation (DG). All this has major potential implications for utilities, new business models, and consumers. A major policy driver in these developments is the need to transition the global economy to a low-carbon economy which requires the massive increase of the proportion of RES or other low-carbon power sources into the energy mix. This transition requires the development and implementation of a smart grid infrastructure. As the RES proportion in the energy mix rises, there is

Power Grid Operation in a Market Environment: Economic Efficiency and Risk Mitigation, First Edition.
Edited by Hong Chen.

also an increased value associated with system and capacity flexibility, in order for the wholesale markets to cope with periods of low renewable output. Changes in the market architecture, introduction of flexible energy market products and installation of flexible assets can certainly aid in this respect. The development of the smart grid infrastructure is also an integral part of the solution to this major challenge.

One of the key questions for the Grid of Tomorrow is how will independent system operators (ISOs)/transmission system operators (TSOs), distribution system operators (DSOs), and aggregators leverage end-to-end machine communications to assure system balance and service reliability at the lowest possible cost to consumers. The prevailing industry view as we move forward into the future is that it will entail (a) broadcasting of dynamic prices and (b) telemetry backhaul to market participants. It is assumed that system control will be then simply an emergent feature of a competitive market design. DR and DER aggregators are often regarded as transaction brokers between end customers and various upstream market participants. The "market failure-free" design for a pure market-driven solution under this paradigm has been elusive, despite decades of research and development. Market failures have been compensated for by repeated regulatory intervention, rather than by an innovation-based engineering solution.

Furthermore, the concept of the ISO/TSO evolved in the era of centralized power generation. The rise of DER, ESR, and DR technologies will require rethinking of how we define ISO/TSO power markets and modifications to market rules in order to accommodate these resources.

The ubiquity of Internet communications itself now permits us to consider using the Internet platform itself for end-to-end communications between machines that produce and consume electricity, automating humans out of the loop. This is one of the opportunities promised by "the IoTs." The control approach we wish to investigate is inspired by the very automatic cooperative protocols that govern Internet communications (the Internet Protocol Suite). These protocols represent a distributed and federated control paradigm, in which information and decision-making authority remain local, yet system stability as a whole is ensured.

DR and DER are two sides of the same coin: resources that are located at the edge of the network, in close proximity to end-user consumption, and largely equivalent from a system standpoint. Arguments can be made for treating DR and DER as either a system supply resource or equivalently a load-shaping technology for load-serving entities. The aggregator of the future may aggregate both types of resources to offer balancing, energy, and DR services to the system. With the increased penetration of renewable energy and accelerated generating plant retirements due to grid decarbonization efforts, DR resources can play an increasing role in balancing the system. Further, as rooftop solar installations increase exponentially, the traditional net demand curve is drastically changing. There is no better demonstration of that than the California ISO's "duck chart," shown in Figure 3.1, which plots the net demand (demand minus renewables) in California and how it is forecasted to change over the next coming years. The belly of the duck gets progressively lower during the day when solar energy is produced, resulting in very steep ramps at sunset and potential overgeneration conditions in early afternoon. These conditions create a significant

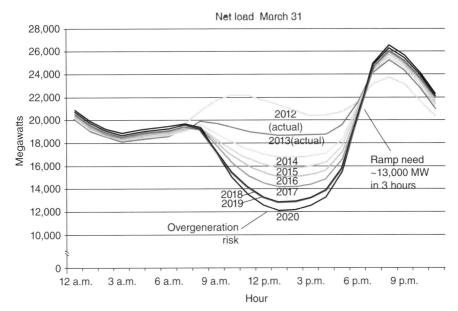

Figure 3.1 The California "duck chart."

challenge for ISOs/TSOs exerting significant stress on the grid and posing serious reliability concerns.

DR resources can be very effective in mitigating this problem by effectively shifting the net demand, elevating the belly of the duck and reducing the morning and evening onramp. This can be achieved by increasing demand during overgeneration conditions and shaving demand during the early night peak. The key challenge we face is how to best utilize these resources, overcome the current regulatory and technology problems, and integrate them into the wholesale energy markets of the ISOs/TSOs.

The smart grid infrastructure and the IoTs offer such a promise. Specifically, computer science research has yielded robust insight into *queuing theory*, which enables oversubscribed computing resources to be shared by multiple consumers. The IoTs will permit us to apply the past one hundred years of advances in queuing theory to the Grid of Tomorrow, by subtly shifting the schedules of billions of devices in real time based on the conditions of the grid, that is, the "grid weather." There are now many demonstrated applications of stochastic control techniques that are surprisingly simple in concept, but field-proven to handle control problems of great complexity. These include the CSMA/CD algorithms that permit cellular phones to share narrow radio frequency bands, telephone switch control algorithms and operating system thread scheduling, as well as examples from nature such as social insect hive behaviors and bacterial quorum sensing.

However, despite these advances, the current application of DR technologies in the field and their impact on market and system operations has been less than encouraging. DR, particularly when price-driven, suffers from dramatic volatile shifting of demand from one time period to another. DERs, such as solar and wind, have similar issues due to weather volatility. This can affect system stability. An approach is emerging where we can moderate the collective ramping behavior of large populations of devices by use of stochastic control techniques. The development and demonstration of this approach, which is a type of priority service pricing, will be the focus of this chapter.

Our proposed architectural approach supplements the inadequacy of pure market-based control approaches by introducing automated, distributed, and cooperative communications feedback loop between the system and large populations of cooperative devices at the edge of the network. ISO/TSO markets and the evolving DSO markets of the future will have both deep markets and distributed control architecture pushed out to the edge of the network. This smart grid architecture for DR in the mass market is expected to be a key asset in addressing the challenges of RES integration and the transition to a low-carbon economy.

3.2 INTRODUCTION

It has been well understood that participation of customers in the form of DR in wholesale organized energy markets, administered by ISOs/TSOs, helps increase the competition and improve the efficiency of these markets. The benefits from DR participation in wholesale energy markets are substantial: (a) it can assist ISOs/TSOs to balance supply and demand by providing an alternative mechanism to dispatching high-supply resources to meet demand, thus reducing market prices and market volatility [1]; (b) it mitigates supply market power by increasing competition and putting downward pressure on generators' strategies to bid at high levels; and (c) it enhances system reliability and resource adequacy because DR resources can provide fast balancing of the transmission grid in case of loss of generation or other unexpected events [2].

Regulators and policy makers have taken substantial measures over the years to encourage participation of DR in wholesale energy markets. For example, in the United States, Congress has instituted national policy to enhance DR participation [3] and FERC has issued over the last several years a series of Orders, including Order No. 890, Order No. 890-A, Order No. 719, and Order No. 745 for the same purpose.

In response to these Orders, several RTOs and ISOs have instituted various types of DR programs. Some of these programs are designed to respond to reliability and emergency conditions, while others are designed to allow wholesale customers, qualifying large retail customers and aggregators, or curtailment service providers (CSPs) of retail customers to participate directly in the day-ahead market (DAM) and real-time energy market (RTM), certain ancillary service (AS) markets, and capacity markets.

Despite these efforts, DR participation has been less than encouraging. There are serious market and regulatory obstacles, and the incentives of utilities and the

CSPs are not well aligned. Utilities until recently have emphasized emergency programs with high-capacity payments to attract customers and have more emergency DR than needed by the RTOs/ISOs. Moving customers to "earlier," more frequent or price-based triggers has not worked as expected. Many states prohibit default dynamic pricing for residential customers. As a result, it is difficult to align pricing for generation, load, and DR resources; this misalignment can lead to gaming opportunities, cost shifting, and perverse incentives.

Regulation and spinning reserve AS markets revenues are essential to DR but are precluded by certain reliability councils, such as the Western Electricity Coordinating Council (WECC). In addition, utilities are perceived to have significant advantages in terms of market power, influence with regulators, and ability to shift administrative costs. State-approved investor-owned utility DR programs reflect multiple policy priorities and program costs may not be fully recovered in wholesale markets. If costs cannot be recovered from the market, either the market structure is wrong or DR is not as valuable as current payments suggest.

In addition, infrastructure and technological barriers still exist which prevent massive penetration of DR resources into the market. There are also market gaming concerns due to "baseline" issues, arbitrage between zonal and nodal pricing, double counting, selective bidding, and so on. Actual curtailment of demand can only be estimated. It cannot be measured and it is grossly imprecise and inaccurate. As a result, even though DR resources should be treated exactly comparable with supply resources, ISO/TSO operators cannot fully depend on them at the present time.

Finally, current DR programs are difficult to understand, require sophisticated energy customers, and are inherently coercive as they are based on either direct load control or punitive dynamic rate structures. Retail customers prefer simplicity and consistency and these preferences are in conflict with the volatility of the wholesale markets. This is consistent with the experience from other markets. For example, for decades telecommunication services were priced using time-of-use per-minute/per-KB rates, ostensibly to manage peak congestion. After deregulation and the resulting disruptive innovations (including cooperative congestion management, i.e., TCP/IP), the vast majority of customers today opt to purchase telecommunication services as a usage-tier-based monthly service rather than as a commodity. Free market choice does not necessarily imply forced participation in government-supervised centralized auction markets; it is also about choice of market venue participation and different value-exchange models.

Indeed, a significant body of research shows that real-world customers do not act as so-called "rational economic agents" who are continuously pondering their marginal consumption utility versus the real-time market price [4]. For ongoing expenses (e.g., electricity, Internet service, cell phones, rent) consumers, in general, strongly prefer fixed-cost "level of service" plans over uncertain dynamically determined costs, even when the uncertain pricing is expected to result in some savings [4]. In effect, consumers are trading a slightly higher price for a lower risk and a lower cognitive burden. Thus, for example, broadband Internet providers offer consumer service classes based on advertised speed and cell phone providers offer service classes based on coverage, minutes-per-month, and free calling groups.

In this chapter, we will present a fundamentally different approach to the DR problem. It is based on distributed computing in a smart grid architecture and is ideal for self-organization of loosely coordinated devices present in micro grids and autonomous systems [5, 6]. It solves many of the problems discussed in this section; it is robust and accurate, very fast and scalable, and not coercive. It is voluntary and bottom up and as such it does not suffer from the severity of the privacy concerns that DR programs currently encounter.

3.3 DISTRIBUTED COMPUTING-BASED DEMAND RESPONSE MANAGEMENT APPROACH

In the proposed approach the residential customers are not directly involved in auction markets and the coordination problem of determining which devices should consume power at what times is solved through distributed aggregation and stochastic control. The proposed method is called ColorPower. The algorithmic challenges of coordinated decision-making in a system of thousands to millions of devices are presented in [5, 6]. In this chapter, we will provide an overview of the proposed distributed computing-based methodology and will address the challenging problem of integrating the ColorPower methodology with energy markets.

At its root, the ColorPower approach is based on the fact that residential and small-business consumers have a great deal of flexibility in their energy needs and are willing and able to adjust levels of demand, *if doing so can be made convenient for them*. Under the ColorPower approach, the consumer designates devices or device modes using "colors" that correspond to control plans, governed by an external or internal load control module that is in communication with a master utility or micro-grid controller (the ColorPower controller). These "color" markings are then used to organize devices into priority groups, dictating when devices are available for demand shaping and the order in which classes of devices will be enabled or disabled. Customers will typically select devices that autonomously cycle off and on (such as HVACs, pool pumps, refrigeration, and water heaters) or those that can gracefully degrade the level of service (such as commercial or public lighting and variable speed motors). Serendipitously, these tend to be larger loads than the devices that customers expect instantaneous noncurtailable service from devices such as residential lighting, clock radios, and other small electrics. We thereby relieve the customer from the annoyance of having to micromanage their net premise power usage in response to grid events.

The ColorPower algorithm operates by aggregating this demand flexibility information into a global estimate of total consumer flexibility. For each color, the algorithm tracks the total demand of devices under management, how much of that demand is currently enabled, and how much demand could be enabled or disabled immediately. The algorithm provides for a "refractory" state (see next section) that avoids fast switching of devices between enabled and disabled, in order to protect equipment and avoid annoying consumers. Aggregation also has the beneficial side effect of preserving the privacy of individual consumers: their demand information simply becomes part of an overall statistic. This aggregate model and the current demand target are then broadcast via IP multicast throughout the system, and every local controller (typically one per consumer or one per device) combines the overall model and its local state to make a stochastic control decision. Given the distributed configuration of the proposed approach, the bandwidth requirements for sending and receiving information is minimal and clearly does not pose a barrier to entry. With each iteration of aggregation, broadcast, and control, the overall system moves toward the target demand, set by the utility or the ISO/TSO, allowing the system as a whole to rapidly achieve any given target demand and closely track target ramps [5,6]. Thus, from a technical perspective, ColorPower is capable of solving the distributed demand management problem for systems of millions of devices.

Now let us consider the consumer pricing perspective. Since we do not require dynamic pricing in order to achieve control, we have a great deal of freedom in the design of consumer pricing models. Indeed, there is no reason that ColorPower could not be connected to a dynamic pricing model if that were determined to be the best way to get consumer adoption. As we discussed in the previous section, consumers, in general, strongly prefer ongoing fixed-cost "level of service" plans over uncertain dynamically determined costs, even when the uncertain pricing is expected to result in some savings.

With the ColorPower approach, these sorts of "service-level" plans are easy to design and implement: we can simply map the "color" designations of electrical devices to plans. By designating a set of devices with particular colors, a consumer is opting into a certain mixture of plans. The only information the consumer needs is the meaning of each color. From a pricing discrimination perspective, what a consumer now is choosing is the level of electrical *time-inflexibility* that they wish to pay for. Participation in a ColorPower program will result in discounted electricity rates that can be designed and implemented by the regulator. The consumer determines the level of its participation by the choice of its service-level plan (i.e., the color designations). A "more flexible" color means less certainty of when a device will run (e.g., time when a pool pump runs) or lower quality service delivered by a device (e.g., wider temperature ranges, slower electrical vehicle charging). These types of economic decision-making are eminently compatible with consumer desires and economic design, as evidenced by the wide range of quality-of-service contracts offered in other industries. This fundamental property of the ColorPower approach can bypass problems, discussed in Section 3.2, related to regulatory biases against dynamic pricing mechanisms and so on.

3.4 THE COLORPOWER ARCHITECTURE AND CONTROL ALGORITHMS

The general ColorPower architecture, (first proposed in [7] and further developed in [5, 6, 8]), automatically matches the demand-shaping requests of a utility or an ISO/TSO with customer preferences for flexibility. Customers mark devices with a

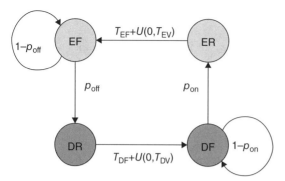

Figure 3.2   Markov model-based device state switching.

qualitative flexibility "color." For example, "green" devices can be shut off at any time, "yellow" only at peak power, "red" in emergencies, and "black" can never be shut off. The customer also has the ability to temporarily override at any time and exit the program. The proposed approach is based on the fact that the system has sufficient flexibility and the regulator will provide sufficient incentives, through discount rates, to ensure consumer participation and investment recovery. Local measurements will be aggregated to form a shared summary model that will be used for distributed shaping of demand. ColorPower devices switch between states according to the modified Markov model, shown in Figure 3.2, between Enabled (E) and Disabled (D) states probabilistically and from Refractory (R) to Flexible (F) states by a randomized timeout.

The block diagram of the ColorPower control architecture is given in Figure 3.3. Each ColorPower client (i.e., the controller inside a device) regulates the

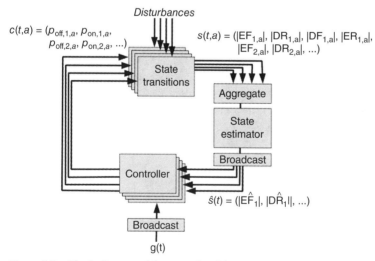

Figure 3.3   Block diagram of the control architecture.

state transitions of the devices under its control. Each client state $s(t,a)$ is aggregated to produce a global state estimate, which is broadcast along with a goal $g(t)$ (the demand target set by the utility or the ISO/TSO), allowing clients to shape demand by independently computing the control state $c(t,a)$. In the remainder of this section, we review (a) the control problem, (b) the control algorithm, and (c) the effectiveness of the control algorithm.

3.4.1 The ColorPower Control Problem

The control problem for a ColorPower system is shown by the block diagram in Figure 3.3: a set of n clients, each controlling a set of electrical devices organized into k colors, where lower numbered colors are intended to be shut off first (e.g., 1 for "green" pool pumps, 2 for "green" HVAC, 3 for "yellow" pool pumps) and where each tier has its own time constants.

Within each color, every device is either *Enabled*, meaning that it can draw power freely, or *Disabled*, meaning that it has been shut off or placed in a lower power mode. In order to prevent damage to appliances and/or customer annoyance, devices must wait through a *Refractory* period after switching between *Disabled* and *Enabled* before they return to being *Flexible* and can switch again. These combinations give four device states (e.g., *Enabled* and *Flexible* (*EF*)), through which each device moves according to a modified Markov model as shown in Figure 3.2: randomly from *EF* to *DR* and *DF* to *ER* (becoming disabled with probability p_{off} and enabled with probability p_{on}) and by randomized timeout from *ER* to *EF* and *DR* to *DF* (a fixed length of T_{*F} plus a uniform random addition of up to T_{*V}).

The state $s(t,a)$ of a client a at time t sums the power demands of the device(s) under its control, and these values are aggregated using a distributed algorithm (e.g., a spanning tree in [6]) and fed to a state estimator to get an overall estimate of the true state $s(t)$ of total demand in each state for each color. This estimate is then broadcast to all clients (e.g., by gossip-like diffusion in [6]), along with the demand-shaping goal $g(t)$ for the next total *Enabled* demand over all colors. Finally, the controller at each client a sets its control state $c(t,a)$, defined as the set of transition probabilities $p_{on,i,a}$ and $p_{off,i,a}$ for each color i. Finally, demands move through their states according to these transition probabilities, subject to exogenous disturbances such as changes in demand due to customer override, changing environmental conditions, imprecision in measurement, and so on.

Note that the aggregation and broadcast algorithms must be chosen carefully to ensure that the communication requirements are lightweight enough to allow control rounds a few seconds long on low-cost hardware. The choice of algorithm depends on the network structure: for mesh networks, for example, spanning tree aggregation and gossip-based broadcast are fast and efficient (for details, see [6]). In general, however, the system must use aggregation and broadcast because it is not currently economically feasible to deploy reliable communications for load control devices with enough bandwidth to poll individual state and deliver individual instructions for millions of devices every few seconds. Aggregation reduces bandwidth requirements by compressing the state of many devices into a single "collective state" message to a central system, while broadcast reduces bandwidth requirements by sharing a single "collective control" message from the central system with many devices.

3.4.2   ColorPower Control Algorithm

The ColorPower control algorithm, determines the control vector $c(t,a)$ by a stochastic controller formulated to satisfy four constraints:

Goal tracking: The total *Enabled* demand in $s(t)$ should track $g(t)$ as closely as possible: that is, the sum of *Enabled* demand over all colors i should be equal to the goal. This is formalized as the equation:

$$g(t) = \sum_i |EF_i| + |ER_i|$$

Color priority: Devices with lower numbered colors should be shut off before devices with higher numbered colors. This is formalized as

$$|EF_i| + |ER_i| = \begin{cases} D_i - D_{i+1} & \text{if } D_i \leq g(t) \\ g(t) - D_{i+1} & \text{if } D_{i+1} \leq g(t) < D_i \\ 0 & \text{otherwise} \end{cases}$$

so that devices are *Enabled* from the highest color downward, where D_i is the demand for the ith color and above:

$$D_i = \sum_{j \geq i} |EF_j| + |ER_j| + |DF_j| + |DR_j|$$

Fairness: When the goal leads to some devices with a particular color being *Enabled* and other devices with that color being *Disabled*, each device has the same expected likelihood of being *Disabled*. This means that the control state is identical for every client.

Cycling: Devices within a color trade off which devices are *Enabled* and which are *Disabled* such that no device is unfairly burdened by initial bad luck. This is ensured by asserting the constraint:

$$(|EF_i| > 0) \cap (|DF_i| > 0) \Rightarrow (p_{on,a,i} > 0) \cap (p_{off,a,i} > 0)$$

This means that any color with a mixture of *Enabled* and *Disabled Flexible* devices will always be switching the state of some devices. For this last constraint, there is a tradeoff between how quickly devices cycle and how much flexibility is held in reserve for future goal tracking; we balance these with a target ratio f of the minimum ratio between pairs of corresponding *Flexible* and *Refractory* states.

This means that in the boundary tier b, whenever some enabled devices are flexible and some disabled devices are flexible, these devices always have a chance of changing their state. For this last constraint, there is a tradeoff between how quickly devices cycle and how much flexibility is held in reserve for future goal tracking; we balance these with a target ratio f, which specifies a lower bound on the permitted ratio of flexible to refractory power.

Since the controller acts indirectly, by manipulating the p_{on} and p_{off} transition probabilities of devices, the only resource available for meeting these constraints is the demand in the flexible states EF and DF for each tier. When it is not possible to satisfy all four constraints simultaneously, the ColorPower controller prioritizes the constraints in order of their importance. Fairness and qualitative color guarantees are given highest priority, since these are part of the contract with customers: fairness by ensuring that the expected enablement fraction of each device is equivalent (though particular clients may achieve this in different ways, depending on their type and customer settings). Qualitative priority is handled by rules that prohibit flexibility from being considered by the controller outside of contractually allowable circumstances.

The other three constraints are satisfied by treating the available demand flexibility as a "budget" and allocating flexibility to each constraint in turn until either all constraints are satisfied or all flexibility is spent or reserved for possible future needs. First comes goal tracking, the actual shaping of demand to meet power schedules. Second is the soft color priority, which ensures that in those transient situations when goal tracking causes some devices to be in the wrong state, it is eventually corrected. Cycling is last, because it is defined only over long periods of time and thus is the least time critical to satisfy.

Each client independently computes a probability distribution over possible actions by bringing all of these constraints together with its own information about device type and customer settings. Each client then samples its distribution and adjusts the state of the devices it controls accordingly, resulting in an aggregate system behavior expected to produce near-optimal system behavior.

3.4.3 Effectiveness of the ColorPower Control Algorithm

The convergence and resilience of the ColorPower control algorithm has been evaluated analytically and empirically in [5]. The analysis focused on the treatment of the case of rapid change of demand goal, for example, following a major equipment failure or renewable generation fluctuation. In this case, the expected number of control rounds r_c for the controller to converge to within ε watts of the goal is logarithmic:

$$r_c = \frac{\log \varepsilon - \log \Delta}{\log(1 - \alpha)}$$

(where Δ is the initial watts of distance from the goal and α is the proportionality of error that goal tracking attempts to correct in each round), provided that a flexible reserve of at least $1/(2+2f)$ is maintained within any given color. Thus, for example, the expected time to converge to within 1% of the goal given 10-s rounds and $\alpha = 0.8$ would be 30 s—quite fast enough to handle most rapid shedding conditions.

The time for the ColorPower controller to return to a quiescent state, from which another such rapid shift is guaranteed to succeed, is conservatively estimated to be equal to

$$r_q = r_c + r_p + 2(T_{DF,b'} + T_{DV,b'}/2 + T_{EF,b'} + T_{EV,b'}/2)$$

where b and b' are the colors on the boundary between *Enabled* and *Disabled* before and after the goal change, respectively, and where r_p is equal to

$$r_p = \max(T_{DF,b'} + T_{DV,b'}, \ T_{EF,b} + T_{EV,b})$$

Simulation studies in [5] confirm these results and demonstrate that the controller scales well on a range of at least 1000–1,000,000 devices that it can closely track ramps even when the ramp direction changes frequently, and that control is robust against changing device populations, noise in estimates, and high degrees of heterogeneity between devices.

3.5 INTEGRATION WITH THE WHOLESALE ENERGY MARKET

In this section we present a methodology for integrating the DR capacity created by the ColorPower system with the wholesale energy markets. Our objective is to create "price-responsive retail demand curves without 'price signals.' " As discussed in Section 3.1, the vast majority of retail customers today opt to purchase services as a usage-tier-based monthly service rather than as a commodity [4]. They opt for simplicity and predictability. Clearly dynamic pricing can be accommodated with this methodology but it is not a substitute. In this section we present (a) the problems of current integration methods and (b) a new methodology for integrating the DR capacity into wholesale energy markets.

3.5.1 Problems with Current Integration Methods

It is generally believed that smart meter deployment and dynamic pricing will result in a retail auction-based solution where consumers will be forced to act as rational economic agents and respond to changing prices. It is envisioned that when this becomes too burdensome, electrical device manufacturers will create devices with energy market trading intelligence; these devices will be programmed by consumers with price ranges to control the maximum price at which they will operate. This widely held view contradicts what residential customers really want: a usage-tier-based monthly service [4].

Good references that summarize the key obstacles to dynamic pricing of electricity on the retail level are in [9–12].

Current DR approaches treat as a supply resource "virtual power plants or proxy generator resources" supplying "negawatts," where bids represent increasing levels of inconvenience with corresponding monotonically decreasing prices at which retail customers are willing to forego consumption. In this model DR may participate in the

DAM including the reliability unit commitment (RUC), the hourly or 5-min RTM, and the day-ahead or real-time AS markets.

This general approach may be practical for large industrial customers but it creates several problems for retail residential customers that severely limit the value of this approach. For example, the management of the data required for scheduling residential customers in zonal level is very taxing since it requires aggregation of numerous very small end-use customers, with frequent migration, that is, enrollments and de-enrollments in a DR program. Further, the use of standard historical load distribution factors (LDFs) that are derived from the EMS State Estimator may also be problematic.

There are also serious settlement concerns. Setting aside the complexity of the settlements of the curtailed portion of the load with the aggregators and the rest of the load with the LSE and the subsequent bilateral settlement between the aggregator and the LSE outside of the ISO's/TSO's settlement process, the determination of the actual DR delivery is very problematic. It is expected to be derived from measurement of aggregate meter usage, calculated from a predetermined, administrative set baseline. The baseline problem, that is, the challenge of measuring what the consumer *would have* done without the payment, has proven to be in practice an intractable problem that has resulted in limited participation of active DR and, even worse, ultimately unreliable DR resources. Finally, the gaming opportunities that result from the baseline problem can be severe. A comprehensive treatment of this gaming problem is contained in [13]. The verification of the performance against the baseline on an aggregate basis rather than by calculating each end-use customer's baseline versus actual and summing the results is also problematic.

There are two additional problems from current practices that are also worth noticing. The first problem relates to the potential double payment for DR that arises if the aggregator or the CSP receives a payment that exceeds the difference between the locational marginal price (LMP) and the implied wholesale price in the current retail price. Such double payments increase rates and could incent inefficient DR whose costs to the market exceed its benefits. The second problem is related to the potential arbitrage between a low load zonal price used to settle the DAM load schedule and the higher nodal price used to settle the curtailed demand. This problem can be limited in magnitude, but not eliminated, by administrative rules such as constraining the number of hours that a DR resource can be dispatched and by requiring the presence of physical control devices. However, adoption of such rules can severely limit the long-run potential of active DR participation in the wholesale market.

3.5.2 New Proposed Integration Method

The proposed integration approach solves the problems identified in the previous section. The self-identified priority tiers of the ColorPower approach to DR enable us to propose a new indirect model for retail power participation in energy markets. Since the demand has been differentiated into tiers with a priority order, the demand in each tier may be separately bid into the current market architecture. The adoption of the proposed approach does not require a change in the current market architecture. The price for each tier is set based on the cost of supplying DR from that tier, which in turn is linked to the incentives necessary to secure customer participation. The tiers are then ordered by priority to form a monotonically decreasing demand curve. Demand can thus express preferences before market clearing in the form of *lower buy bids* for lower priority tiers of demand. This allows aggregated demand to send price signals in the form of decreasing buy bid curve. Market information thus flows bidirectionally.

The ColorPower system can be used by a retail utility or an aggregator to bid an aggregate retail demand curve into the ISO/TSO markets on behalf of its customers, neither requiring customers to interact with real-time prices nor submitting a price curve for DR. Customer demand is instead aggregated and represented by a utility procurement trader, who purchases as much power as is deemed rational to purchase, that is, up until supply prices start to rise at levels not acceptable by the purchaser. The device population then shares the limited available power resource, and any shortages are handled in an orderly way by automatically diffusing small curtailment periods across the population. This automatic diffusion of inconvenience across the entire population of devices is the cornerstone capability of the proposed approach.

The ColorPower system permits customers to self-identify devices within their total load profile that are flexible. The level of flexibility each customer selects will determine the applicable discount rate designed by the regulator. This capability effectively creates two tiers of power demand, flexible P_{fn} and inflexible P_i, that can be aggregated and monitored at the utility level. Additional tiers of prioritized flexible demand can be defined and designated as P_{f1}, P_{f2}, and so on. Each tier can be separately forecasted by the utility, and bids can be submitted into the market to procure supply for the entire power demand P_t such that

$$P_t = P_i + (P_{f1} + P_{f2} + \cdots + P_{fn})$$

For customers, the colors correspond to contractual constraints on when resources with that color are available for flexibility. For example, a thermostat controlling a HVAC system might have a "green" range for any-time flexibility and a "yellow" range for flexibility during peak power and emergencies. Demands in tiers whose constraints make them currently unavailable are bid just as unmanaged demand, while bids for potentially flexible demand is selected to maximize profits subject to contractual constraints with customers, just as generators currently bid supply.

These loads are aggregated and can be bid directly into the markets distinct from the traditional uncontrolled inflexible demand as shown in Figure 3.4. Ideally, the inflexible demand component is bid at a price at which the supply price curve is forecasted to be flat. The flexible demand component can be bid near the inflection point of the supply curve. Depending on the market design rules, the buyer's clearing price may be held secret in order to encourage competitive bidding or the buyer may be required to present a matching DR supply bid to the market that represents the buyer's price at which curtailment is substitutable for supply. Designing a mechanism for optimization of such bids is beyond the scope of this chapter, but is expected to follow the lines of prior work on reliability pricing such as that described in [14,15].

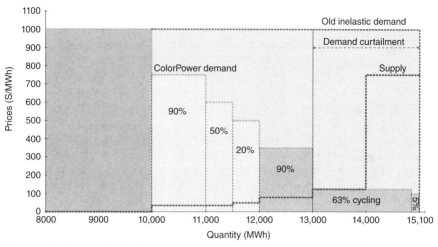

Figure 3.4   Monotonically decreasing demand curve based on aggregated, differentiated tiers of measured customer demand.

Utility buyers make this decision on behalf of aggregated segments of customer demand, taking into consideration customer inconvenience and other factors. Customers signal that their curtailment tolerance has been reached by opting their devices out of the program. A small amount of flexible demand can buffer the volatility of the overall power demand by yielding consumption to the inflexible components as necessary based on the customer-expressed priority, while fairly distributing power to all customer devices within a demand tier.

3.6 EQUALIZING MARKET POWER BETWEEN SUPPLY AND DEMAND

The proposed approach has the potential to transform the power markets by *eliminating irrational marginal power fulfillment at the retail level that is not cost effective to be fulfilled*. Wholesale power markets feature "hockey stick"-shaped supply bids during shortages, due to exploitation of scarcity pricing power or the expense of running generation resources in excess of their most efficient output levels. Both generation and demand benefit from reducing uneconomic energy demand fulfillment.

The monotonically decreasing demand curve greatly reduces the incentive for strategically withholding supply from markets in order to sell it at scarcity pricing levels. The marginal price is now set by flexible demand bids, not supply bids. This may erode the profitability of energy speculation while lowering prices for the public. This demand curve can also reduce the need for financial hedging, by using physical demand reduction to reduce under-capacity risk. Reductions in hedging costs will reduce costs across all power market participants and ideally to the public.

Certain generation resources are often designated as "load-following." With the ColorPower approach, certain load resources can be designated as "supply-following," in particular loads directed to follow intermittent RES. This capability can be used to manage balancing and ramping challenges for massive RES penetration instead of solely relying on expensive conventional flexible generation that requires massive expenditures in infrastructure projects and can result in noncompliance with environmental goals.

Supply-following loads can also be made dispatchable as an opportunistic or emergency power sink. These can be coupled to real-time telemetry signals such as substation SCADA demand or wind farm telemetry. A demand tier consisting of "pool pumps that run only when excess renewable power is available" can wait at a very low buy bid—even at negative bids—for loads that can be run in an emergency to consume excess power. This type of demand can be used to consume excess power from hydro plants or excessively windy days and customers can be paid by generation to consume it.

It should be noted at this point that exposing individual retail customers directly to auction pricing reduces their market power significantly compared to the ColorPower aggregate buying approach. Individual customers are presented with a "take-it-or-leave-it" monopolistic price from their power supplier, and only have the choice to micromanage consumption at the entire premise level. The ColorPower approach suggests a role for utilities in the future as active managers of aggregate demand fulfillment. Incremental costs for an aggregator are much lower due to the economy of scale for one organization to monitor and manage demand delegated to it by large numbers of customers. The utility can then act as an effective negotiator between supply and demand to provide the best overall balance of service levels and cost for their customers, as opposed to being a mere pass-through distributor of a commodity that marks up whatever price is currently being demanded by generation.

3.7 GENERALIZATION BEYOND DEMAND RESPONSE

Although this chapter has focused primarily on DR, the proposed approach has the potential to offer more general demand management services, which can also help manage other resources such as dispatchable demand and distributed supply. Specifically, the same approach demonstrated above can be easily extended to include DER and ESR without the intractable computational problems that nodal settlement would present at the local level. Generation such as solar, wind, combined heat and power (CHP), gas turbine, and other "behind-the-meter" distributed resources can be modeled as DR resources. This section focuses on extending the proposed approach to DERs in ISO/TSO wholesale level. However, similar concepts can apply to the emerging local energy markets at the distribution level (Distribution System Operators or DSOs).

These could be aggregated into economically significant blocks by extending ColorPower to include supply tiers, then bid into the ISO/TSO markets as supply or demand segments to compete against imported power resources. If configured as a supply resource, they can bid against other supply resources normally; if configured as a demand resource, they can be interleaved with DR resources to create mixed dispatch orders, for example, a generation or storage resource might be in a priority tier dispatched before more disruptive emergency DR resources are used.

Storage resources combine characteristics of dispatchable demand and generation: their demand or supply power corresponds to their rate of discharge or charge, while the length of time each resource may be available depends on the amount of energy currently stored. Storage resources can thus be aggregated using ColorPower and bid in at opportunistic prices, like dispatchable demand, to absorb excess supply and assigned to an appropriately priced priority tier to determine when they discharge. This can be further refined by assigning different levels of storage to different priority tiers, so that, for example, a storage resource might always maintain a small operating reserve to be sold into an AS market.

Extending the proposed approach to offer more general demand management services requires certain changes in the structure of the current ISO/TSO market architecture. In today's markets, centralized generators have a *resource identity* (resource ID) and are required to have "revenue quality metering." It can be via a direct interaction between the ISO/TSO and the resource ID, or it can be through a certified market participant, like a scheduling coordinator, that mediates between the ISO/TSO and the resource ID. But for distributed resources, assigning a resource ID to each resource is not feasible because the cost would be prohibitive. The solution to this problem requires the certified market participant, like a scheduling coordinator, to take administrative control of the aggregated distributed energy accounts and meter them with any technology, including any online technology, that suits their purposes. The aggregator can be its own certified market participant or can hire a third party. A directly connected interface between the ISO/TSO and the aggregator is no longer required, and any communication network or protocol that provides the necessary data in a timely way should be acceptable.

The aggregation of distributed resources needs to follow certain rules: if they span multiple pricing points they have to be of the same type and they have to act in the same manner, that is, they should all be injecting energy into the grid or ejecting energy from the grid. In other words, all sub-resources must be homogenous and they must move in the same direction as the ISO/TSO dispatch instruction. Homogenous aggregations are those in which all sub-resources are generation, energy storage acting together in charge or discharge only, or are load. For aggregations of energy storage, all sub-resources must be operating in the same mode (i.e., charging or discharging, but not a mix of the two) in response to an ISO/TSO dispatch.

Sub-resources in an aggregation across multiple pricing points can cause distribution variability and this can have negative effects on congestion management. Relaxation of this requirement can be achieved as more experience is gained with the impact of this distribution variability on congestion management. This is especially relevant to aggregated solar-plus-storage technologies that might be producing both load and generation. Clearly it would be desirable if aggregating mixtures of rooftop solar, energy storage, plug-in electric vehicles, and DR across multiple pricing points, without all the proposed limitations, is possible.

Furthermore, the aggregation of distributed resources should be limited to a single zone from congestion viewpoint. The reason is that it is critical to be able to assess their impact on congestion and identify critical constraints. Otherwise, if in a market run a constraint is identified between these two zones, then the specific aggregation would be simultaneously on the "right" side and on the "wrong" side of the constraint. In this case the potential exists that a dispatch instruction issued to this aggregation to alleviate a constraint between these two zones may actually exacerbate the problem.

Storage resources, in particular, combine dispatchable demand and dispatchable generation characteristics. These can be bid in at opportunistic prices to absorb excess supply and discharged at an appropriate place in the ColorPower dispatch order. The appropriate model applicable for ESR assets is the proposed nongeneration resource (NGR) model, as shown in Figure 3.5. NGR is a generic resource model where a resource can produce or consume energy within a continuous operating range that can span both generation and demand. This model can be used for DR, when the operating range spans demand only, or for an ESR that can operate continuously in charging and discharging modes. Figure 3.5 shows a typical energy bid from an ESR.

With the NGR model, an ESR can participate in the following AS markets: (a) regulation down, (b) regulation up, (c) spinning reserve, (d) nonspinning reserves, (e) flexible ramping up, (f) flexible ramping down, and (g) primary reserve.

Regulation up and down awards are remunerated at the corresponding marginal prices and there may be additional payments for regulation up and down mileage (usage), if this feature is included in the market design. Spinning and nonspinning

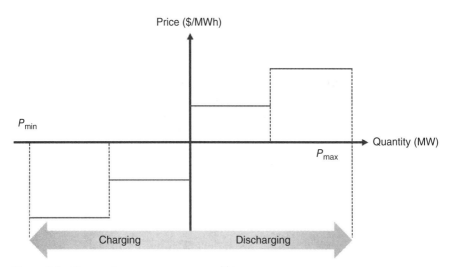

Figure 3.5   Energy storage resource energy bid.

reserve awards are remunerated at the corresponding marginal prices. Flexible ramping up and down awards are remunerated at the corresponding marginal prices, which for these services are driven solely by opportunity costs since normally no bids are allowed for them. Primary reserve, also known as frequency response, is usually not procured as a separate AS; it is traditionally being provided by the governors of large generating units as part of their generation interconnection agreement with the ISO/TSO. However, with the increased penetration of renewable energy and accelerated generating plant retirements due to grid decarbonization efforts, it will not be long before the ISOs/TSOs will have to procure primary reserve and remunerate it at its marginal price, like all other capacity ASs.

ESRs are uniquely qualified to provide all these ASs, especially frequency response and regulation, because they possess superior ramp rates. ESRs may also earn revenues from participating in the energy market. In that respect, they can provide a very beneficial function to the grid by pairing the intermittent resources. For example, market participants may pair their wind energy portfolios with ESR to reduce the volatility of intermittent energy production and earn higher capacity credits.

3.8 A NUMERICAL EXAMPLE

We have implemented and tested the proposed DR approach into the ColorPower software platform. This section presents a numerical example using the assumptions stated in Section 3.3 and the ColorPower software platform. Except where noted otherwise, all simulations are executed with the following parameters (which are the same as for Figure 3.3): 10 trials per condition for 10,000 controllable devices; each device consumes 1 kW of power (for a total of 10 MW demand); devices are 20% green, 50% yellow, and 30% red; the measurement error is $\varepsilon = 0.1\%$ (0.001); the rounds are 10 s long; all the refractory time variables are 40 rounds; the flexible reserve ratio is $f = 1$; and the proportional control constant is $\alpha = 0.8$. Error is measured by taking the ratio of the difference of a state from optimal versus the total power.

To validate our convergence and quiescence time predictions, we consider steps where the goal before and after the step ranges from 0 to 10 MW in increments of 0.1 MW, executing one trial per condition. Simulation begins in steady state at the prior level and proceeds for 3600 s following the step. We consider the controller to have converged to the goal when the Enabled demand never again departs by more than 1% (0.1 MW) average over a minute, and to have reached quiescence when it never departs from the steady-state distribution of demand into states by more than 2% (0.2 MW).

The results of the simulation test are illustrated in Figure 3.6. The figure illustrates a peak shaving case where a power quota, the DR target that may be provided from an externally generated demand forecast, is used as a guide for the demand to follow. This forecast could be the result of an energy simulation of the DAM that determines the impact of certain DR scenarios on market clearing LMPs. When peak

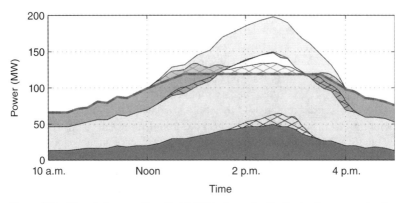

Figure 3.6   Simulation results with 10,000 independently fluctuating power loads.

control is desired, the aggregate demand remains below the quota, while individual loads are subjected stochastically to brief curtailments. Post-event rush-in, a potentially severe problem for both traditional DR and price signal-based control systems, is also managed gracefully due to the specific design of the modified Markov model.

Taken together, these results indicate that the ColorPower approach, when coupled with an appropriate controller, should have the technological capability to flexibly and resiliently shape demand in most practical deployment scenarios.

3.9 CONCLUDING REMARKS

The current status of the DR programs is not satisfactory due to a series of factors including market, regulatory, infrastructure, and technology problems. Most are based on direct control of devices by utilities that allow customers to trade inconvenience for cash. Further, the DR products that ISOs/TSOs are offering are not popular due to a variety of reasons.

This chapter presents a fundamentally different approach to the DR problem. It is based on a distributed computing methodology and is applied to the mass market of residential and small-business customers. The distributed approach allows for scalability, privacy, accuracy, and robust response. It achieves the same economic, reliability, and balancing goals as dynamic pricing and traditional DR, with superior results, without their drawbacks. The chapter presents the proposed methodology and addresses the challenging problem of integrating the methodology with energy markets. Simulation results with 10,000 devices were also presented. The simulation results give credence to the claim that the proposed approach, when coupled with an appropriate controller, has the technological capability to flexibly and resiliently shape demand in most practical deployment scenarios. The capability offered by the proposed methodology is essential to address the challenges we face in integrating massive RES and other distributed assets into the grid toward the transition to a low-carbon economy.

APPENDIX 3.A NOMENCLATURE

Presented here are the key terms and variables for the ColorPower-proposed DR approach.

Symbol	Definition
Color	Customer configuration of quality-of-service plan for an electrical device
Priority tier	Set of devices collected by color and class; lower numbered tiers are shut off first
Enabled	Device or device mode is allowed to consume power
Disabled	Device or device mode is not allowed to consume power
Flexible	Device or device mode can switch between enabled and disabled
Refractory	Device or device mode cannot switch
EF, DR,	States combining enabled, disabled, flexible, and refractory
ER, DF	Fixed rounds of disabled refractory time for tier i
$T_{DF,i}$	Maximum random rounds of disabled refractory time for tier i
$T_{DV,i}$	Fixed rounds of enabled refractory time for tier i
$T_{EF,i}$	Maximum random rounds of enabled refractory time for tier i
$T_{EV,i}$	
t	A point in time
a	A ColorPower agent
$s(t,a)$	State of demand for agent a at time t
$|X_{i,a}|$	Power demand (watts) in state X for color i at agent a
$s(t)$	State of total power demand (watts) at time $s(t)$
$|X_i|$	Total power demand (watts) in state X for color i
$\hat{s}(t)$	Estimate of $s(t)$
$|\hat{X}_i|$	Estimate of $|X|$
$g(t)$	Goal total enabled demand for time t
$c(t,a)$	Control state for agent a at time t
$P_{off,i,a}$	Probability of a flexible color i device disabling at agent a
$P_{on,i,a}$	Probability of a flexible color i device enabling at agent a
D_i	Demand for ith color and above
b	Boundary color, which mixes disabled and enabled demand
f	Target minimum ratio of flexible to refractory demand
$r(t)$	Ratio of the distances from $g(t)$ to the boundary color edges(D_b, D_{b+1})
α	Proportion of goal discrepancy corrected each round
r_c, r_q	Number of rounds for controller to converge/quiesce

REFERENCES

[1] "Demand Resource Compensation in Organized Wholesale Energy Markets," FERC Docket No. RM10-17-000, Order No. 745, March 15, 2011.

[2] National Renewable Energy Lab, "Technical Report TP-500-43373, ERCOT Event on February 26, 2008," July 2008.

[3] Energy Policy Act of 2005, Pub. L. No. 109-58, § 1252(f), 119 Stat. 594, 965, 2005.

[4] H. Mitomo and T. Otsuka, "Consumers' Preference for Flat Rates, A Case of Media Access Fees," International Telecommunications Society, 17th Biennial Conference, 2008.

[5] J. Beal, J. Berliner, and K. Hunter, "Fast Precise Distributed Control for Energy Demand Management," Presented at the 6th IEEE International Conference on Self-Adaptive and Self-Organizing Systems (SASO 2012), September 2012.

[6] V. V. Ranade and J. Beal, "Distributed Control for Small Customer Energy Demand Management," IEEE Self-Adaptive and Self-Organization 2010, September 2010.

[7] J. Beal and H. Abelson, "PACEM: Cooperative Control for Citywide Energy Management," Whitepaper, August 2008.

[8] V. V. Ranade, "Model and Control for Cooperative Energy Management," Master's Thesis, Massachusetts Institute of Technology, June 2010.

[9] S. Borenstein, "Real-time retail electricity prices, theory and practice," in *Electricity Deregulation, Choices and Challenges*, J. M. Griffin and S. L. Puller, Eds., University of Chicago Press, Chicago, 2005, pp. 317–357.

[10] A. Faruqui, "2050: a pricing odyssey," *The Electricity Journal*, vol. 19, pp. 3–11, October 2006.

[11] K. Costello, "An observation on real-time pricing: why practice lags theory," *The Electricity Journal*, vol. 17, pp. 21–25, January–February 2004.

[12] E. Hirst, "Price-responsive demand in wholesale markets: why is so little happening?" *The Electricity Journal*, vol. 14, pp. 25–37, May 2001.

[13] Market Surveillance Committee of the California ISO, "An Analysis of the California ISOs' Demand Response Proposal," May 1, 2009.

[14] H.-P. Chao and R. Wilson, "Priority service: pricing, investment, and market organization," *The American Economic Review*, vol. 77, no. 5, pp. 899–916, December 1987.

[15] S. Oren, S. Smith, and R. Wilson, "Service Design in the Electric Power Industry," Electric Power Research Inst., Palo Alto, CA/Pricing Strategy Associates, Orinda, CA (USA), Technical report, 1990.

CHAPTER 4

IMPROVE SYSTEM PERFORMANCE WITH LARGE-SCALE VARIABLE GENERATION ADDITION

Yuri V. Makarov, Pavel V. Etingov, and Pengwei Du

ELECTRICITY RESTRUCTURING IS ONE of the significant steps that utilities have undertaken toward the efficient and reliable production of electricity. A variety of electricity markets have been proposed and developed. Although these designs vary in features, their fundamental principle is the same, that is, to promote transparent, nondiscriminatory power market for participants. To some extent, the management of those markets has been successful.

However, a future power system with large-scale renewable resources like wind/solar power will create significant challenges to the market operation. The growth of renewable resources is driven by societal benefits, such as low environmental impacts and increased energy security. However, the renewable resources are variable and hard to predict. The prevalent practice is that the wind and solar energy resources are given priority for dispatching against the conventional generators, and independent system operators (ISOs) or transmission system operator (TSOs) are responsible for balancing against their energy. The resulting effect of a large variable generation on system operation could be multifold. First, transmission limits can create significant problems in incorporating wind and solar energy resources into the system. The existing transmission networks are becoming increasingly incapable of delivering the renewable energy to load centers because of transmission constraints. Second, increased operation reserves are needed to compensate for the uncertainty and variability of these resources. This problem is exacerbated when variable generation displaces conventional generators, which provide balancing reserves.

This situation is exacerbated when variable generation displaces conventional generation in the market, which is the predominant resource for operation reserves. For instance, a high price spike due to lack of flexibility has been experienced by California independent system operator CAISO) in June 2011 as shown in Figure 4.1.

Power Grid Operation in a Market Environment: Economic Efficiency and Risk Mitigation, First Edition.
Edited by Hong Chen.

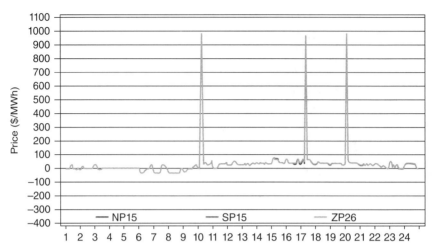

Figure 4.1   CAISO real-time price.

The ancillary service price soared to about $1000 while it was below $100 for normal conditions. When significant departures from the competitive pricing are found, operator interventions are required either to curtail renewable resource or to hold more flexible dispatchable reserves.

Lack of flexibility services available from conventional generators (including start-up time, ramping capability, cycling capability) is considered one of the significant barriers to achieve higher penetration levels of renewable generation. To quantify the flexibility need, this chapter presents several methods to predict the balancing reserve needs and their characteristics, required from conventional generation and other resources, and indicates how to obtain these additional flexibility resources.

4.1   REVIEW OF REGULATION AND ANCILLARY SERVICES

Regulation is the process of providing minute-to-minute system balance by adjusting power output of generating units connected to the automatic generation control (AGC) system. Regulation is typically an expensive ancillary service. Some recent studies on renewable resources indicated that regulation requirements may increase because of the variability and forecast uncertainty associated with the high penetration level of renewable resources such as wind and solar [1–7]. Retirement of conventional generating units is expected within the next decade in some systems. These units have been used to provide balancing services, and their retirement could potentially create some deficiency in available regulation resources. Consequently, the price of regulation would increase because more regulation procurement is needed while available resources providing this service decline.

The power system balancing processes, including scheduling, intra-hour real-time dispatch (load-following), and regulation processes, are traditionally based on deterministic models. These balancing processes work relatively well with predictable traditional generation sources. However, the increasing penetration of the intermittent renewable generation sources, such as wind and solar generation, causes significant challenges in quantifying the uncertainties associated with forecasting intermittent generation resources and in incorporating the uncertainties into grid operation. These uncertainties are not yet reflected in existing energy management systems (EMSs) and generation commitment, dispatch, and market operation tools. With increasing penetration of variable resources, these uncertainties could result in significant sudden unexpected power balancing problems and cause serious risks to system reliability and control performance. Without an assessment of these risks, system operators would have limited information on the likelihood and magnitude of these problems. Furthermore, these uncertainties could require procuring additional costly balancing services. Major unexpected variations in wind power, unfavorably combined with load forecast errors and forced generator outages, could cause significant power mismatches that are unmanageable without knowing these potential variations in advance. There is a need to know whether the system will be able to meet the balancing requirements within the look-ahead horizon, what additional balancing efforts may be needed, and what additional costs will be incurred by those needs.

4.2 DAY-AHEAD REGULATION FORECAST AT CAISO

To achieve a balance between the market efficiency and reliability at CAISO, new methods have been developed, for example, day-ahead regulation requirement forecast, incorporation of the uncertainties caused by variable generation into power system operations, and wide-area energy storage management system. This section presents a new approach, methodology, and software developed at PNNL for procuring regulation capacity that would minimize the required regulation reserve for a particular operating hour of a day without compromising balancing authority's (BA's) control performance characteristics.

The approach is capable of predicting BA's regulation reserve requirement on a day-ahead basis by calculating regulating capacity, ramping rate, and ramp duration requirements, including upward and downward requirements, for each operating hour of a day. Three methods have been implemented. In the first method, the probability distributions of area control errors (ACE) components (including interchange error component, frequency error component, metering error correction component, automatic time error correction component, and inadvertent interchange payback component) are evaluated separately and summed to evaluate the regulation requirement sufficient to address all these uncertainties collectively. The second method predicts regulation requirements based on a statistical analysis of ACE signals and applied regulation data. The third method is similar to the second method, but is based on the new balancing authority area control error limit (BAAL) standard.[1]

A statistical approach based on the time-varying empirical probability density function (PDF) is used in all three methods to determine the regulation requirement. A moving window is defined to collect sufficient statistical information regarding probability distributions of the regulation requirement and its components. Based on the collected statistics, the approaches evaluate the percentile intervals (also called confidence intervals or uncertainty ranges) for each operating hour based on a certain user-specified level of confidence. Actual CAISO operation data were used to validate the performance of all three methods. The results obtained by the three methods are presented and compared. Results show that using the proposed methods could save CAISO an average of 10% of its current regulation procurement, while satisfying previously enforced control performance standards (CPSs). With the new BAAL control standard, the savings could reach 30%.

4.2.1 Methodology

The day-ahead hourly regulation requirements are determined based on a prespecified confidence level.

4.2.1.1 Area Control Error

The equation used by the Western Electricity Coordinating Council (WECC) to calculate ACE is as follows:

$$ACE_{WECC} = (NI_A - NI_S) - 10B_i(F_A - F_S) - T_{0b} + I_{ME} + \frac{II^{on/off\,peak}_{primary}}{(1-Y)H} \qquad (4.1)$$

where NI_A is the actual net interchange (MW), NI_S is the scheduled net interchange (MW), B_i is the frequency bias for the BA (MW/0.1 Hz), F_A is the actual frequency (Hz), F_S is the scheduled frequency (normally 60 Hz), T_{0b} is the bilateral payback for inadvertent interchange (MW), I_{ME} is the metering error correction (MW), B_S is the WECC system frequency bias (MW/0.1 Hz), Y is B_i/B_S, $\frac{II^{on/off\,peak}_{primary}}{(1-Y)H}$ is the BA's accumulated primary inadvertent interchange (MWh), accumulation is calculated separately for on-peak and off-peak hours, and H is the number of hours used to pay back inadvertent interchange energy.

4.2.1.2 Analysis of Forecast and Dispatch Errors (Method I)

The first method uses a statistical analysis of real-time forecast and dispatch errors that affect the CAISO ACE signal to predict day-ahead regulation requirements. The

---

[1] BAAL is designed to replace CPS2 standard [8]. Details on CPS2 can be found in Jaleeli and Van Slyck [9]. The BAAL standard is expected to relax the area regulation needs and reduce the regulation burden on resources providing regulation service.

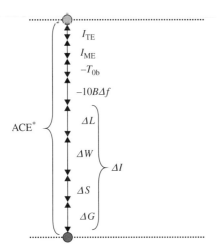

Figure 4.2   Calculated ideal regulation used in method I.

components of the ACE equation are used to calculate 1-min resolution ideal regulation curves that would correspond to zero ACE. This curve consists of the summation of the ACE components listed below, as shown in Figure 4.2.

According to equation (4.1), a BA-ACE includes interchange error component ΔI, frequency error component $10B\Delta f$, metering error correction component I_{ME}, automatic time error correction component I_{TE}, and inadvertent interchange payback component T_{0b}.

Interchange error can be calculated using the equation

$$\Delta I = NI_A - NI_S \tag{4.2}$$

where $NI_A = G_A + W_A + S_A - L_A$, $NI_S = G_S + W_S + S_S - L_S$, G_A is the actual conventional generation (without wind and solar), W_A is the actual wind generation, S_A is the actual solar generation, L_A is the actual load, G_S is the generation dispatch (schedule), W_S is the wind generation forecast (schedule), S_S is the solar generation forecast (schedule), and L_S is the load forecast (schedule). Thus, the interchange error is defined as

$$\Delta I = (G_A + W_A + S_A - L_A) - (G_S + W_S + S_S - L_S)$$
$$= \Delta G + \Delta W + \Delta S - \Delta L \tag{4.3}$$

where ΔG is the uninstructed deviation of conventional generation units, ΔW is the wind generation forecast error, ΔS is the solar generation forecast error, and ΔL is the load forecast error.

If a BA does not have regulation, its ACE would be

$$ACE^* = \Delta G + \Delta W + \Delta S - \Delta L - 10B\Delta f - T_{0b} + I_{ME} + I_{TE} \tag{4.4}$$

Assuming that the BA's goal is to keep its ACE as close as possible to zero, the ideal regulation requirement would be

$$REG^* = -ACE^* \tag{4.5}$$

Method I accumulates and processes statistical information on REG^*, so that the statistical bounds, corresponding to a certain confidence level (for instance, 95%), are determined for the upward and downward regulation requirements. Along with the capacity requirements, ramp and ramp duration requirements are calculated as well.

4.2.1.3   Actual ACE and Regulation Data (Method II)

Instead of using equation (4.4) as in method I, a minute-by-minute regulation can be calculated directly by subtracting the actual ACE from the actual regulation values, as shown in Figure 4.3:

$$REG^* = -ACE^* = REG_A - ACE_{WECC} \tag{4.6}$$

where ACE^* is the ACE that would be observed in the case of no regulation in the BA control area, REG_A is the actual regulation applied to the system, and ACE_{WECC} is the BA-ACE defined in equation (4.1). This REG_t^* represents the regulation needed to have a zero ACE.

Equation (4.6) for the regulation value overestimates the needed regulation because it drives the ACE value to zero at any analyzed moment. This target is neither economical nor practically achievable. A more realistic regulation value can be calculated based on the required compliance level with the CPS requirements. There are multiple possible ways to reduce the regulation reserve capacity from the level required by ideal regulation service to the level needed to meet CPS requirements or any other level desired by the system operator. In this project, an approach was adapted in which the ACE signal is reduced by a value proportional to a BA's L_{10}. The system operator can specify an experimental proportionality coefficient to help reach the desired level of CPS compliance. In this approach, a new term taking into

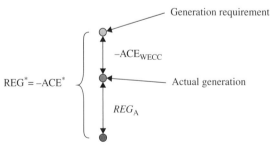

Figure 4.3   Calculated regulation.

account the L_{10} limit [9] can be introduced into equation (4.6) to calculate regulation requirements REG** as follows:

$$REG^{**} = \begin{cases} -ACE^* + \alpha L_{10}, & \text{if } ACE^* \geq L_{10} \\ -ACE^* - \alpha L_{10}, & \text{if } ACE^* \leq -L_{10} \\ -ACE^*, & \text{if } -L_{10} < ACE^* < L_{10} \end{cases} \tag{4.7}$$

where L_{10} is a special limit and the coefficient α can be determined based on the CPS compliance level target. Information on REG** in method II is processed in the same way as in method I.

4.2.1.4 Calculation of Regulation Based on BAAL (Method III)

To meet the BAAL standard requirements, the ACE signal should satisfy the constraint:

$$BAAL_{\min t} < ACE^*_t < BAAL_{\max t} \tag{4.8}$$

where $BAAL_{\max t}$ and $BAAL_{\min t}$ are limits at time t [8]; ACE^*_t is a BA's ACE in the case that the system does not have regulation at time t.

According to the BAAL standard, the ACE signal can be outside limits for a certain time (up to 30 min). A methodology for incorporating the timing component into regulation requirements calculation is given in Makarov et al. [10].

Figure 4.4 illustrates the proposed methodology for identifying regulation needed to meet the BAAL standard requirements. The BAAL limit and the BA-ACE* are shown in the figure. One can see a violation of the BAAL limit from approximately

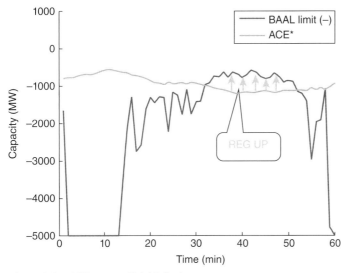

Figure 4.4   ACE* versus BAAL limit.

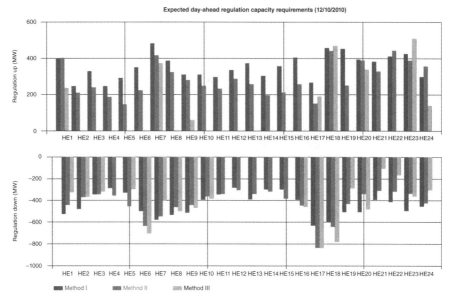

Figure 4.5   CAISO regulation capacity requirements by methods I, II, and III.

30 min to 50 min. To address this violation, a regulation-up control action should be applied (shown as arrows) to keep ACE within the limits.

For each violation, the BAAL allows a BA to have its ACE outside limits for a certain time (up to 30 min). To reflect this flexibility, an additional timing component can be added to the methodology for estimating the regulation needed. Details of this methodology can be found in Makarov et al. [10].

4.2.2   Simulation Results

CAISO actual data have been used in this study. Figure 4.5 shows an example of CAISO's hourly regulation requirement prediction. Prediction results calculated by method I, method II (for the 90% confidence level), and method III are presented.

Ideally, results of applying methods I and II should be close, but nevertheless some difference in the results can be seen; these differences can be explained by the fact that actual CAISO statistical information from multiple data sources has been used in this study. This information contains missing points, outliers, spikes, measurement errors, and so on. To filter the "bad" data, several approaches have been applied; this data preprocessing allows us to get relatively close results.

It can be seen from Figure 4.5 that regulation requirements can be essentially reduced because the BAAL standard allows BAs to operate in a wider range of ACE values and consequently with less regulation compared with the previous CPS.

Figure 4.6 shows the "ideal" regulation requirements curve. This curve corresponds to the regulation that would be needed to have constant zero ACE. The day-ahead regulation actually procured on the same day is shown by the dark gray range,

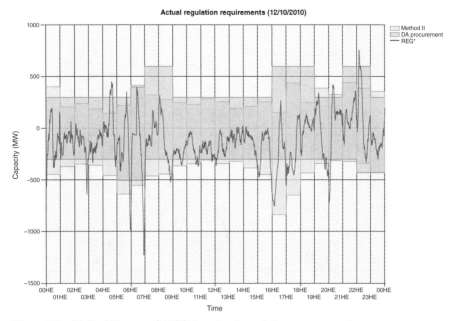

Figure 4.6   Method II versus CAISO day-ahead regulation procurement.

and regulation requirement prediction (for a 90% level of confidence) is shown by the light gray range.

Total day-ahead regulation procurement cost on the analyzed date was more than $112,000. Using the regulation requirements provided by method II would have allowed the CAISO to save about $15,000 (13%) during this day.

An evaluation of the potential monthly savings is shown in Figure 4.7. The total monthly cost of regulation is about $2.5 millions. The average monthly savings for method II is about 11%, or $250,000. Because the new BAAL standard allows a BA to operate in a wider ACE range, using regulation requirements calculated by method III could potentially save as much as 30–40%. AGC algorithms should be adjusted to make use of the relaxed BAAL balancing criteria, so that the actual ACE value does not result in unallowable flows among BAs. Note that the potential saving achieved by method III can be reduced because of other potential implications such as accumulation of inadvertent energy.

4.3 RAMPING AND UNCERTAINTIES EVALUATION AT CAISO

To improve the system control performance, maintain system reliability, and minimize costs related to system balancing functions, it is necessary to incorporate expected wind and load uncertainties into scheduling, load-following, and, to some

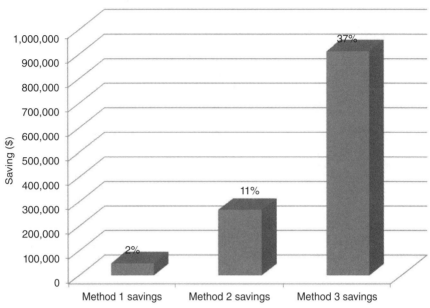

Figure 4.7   Regulation procurement monthly savings under the existing procedure (total cost) achieved by using methods I through III.

extent, regulation processes. Some wind forecast service providers offer uncertainty information for their forecasts. For instance, AWS Truepower [11] and 3TIER [12] developed wind power generation forecasting tools with built-in capability to assess wind generation uncertainty. Similar tools have been developed in Europe. In the context of a European Union project, ANEMOS, a tool for online wind generation uncertainty estimation based on adaptive resampling or quantile regression, has been developed [13]. A German company, Energy & Meteo Systems, developed a tool for wind generation forecasting, assessing the uncertainty ranges associated with wind forecast and predicting extreme ramping events [14]. Pinson et al. [15] discusses a wind generation interval forecast approach using the quantile method. Luig et al. [16] used statistical analysis based on standard deviation to predict wind generation forecast errors. Work is underway to incorporate these uncertainties into power system operations [17, 18]. Wind power generation is taken into account in the unit commitment problem by researchers in [19–21]. Unfortunately, in many cases these efforts are limited to wind generation uncertainties only and ignore additional sources of uncertainty, such as system loads and forced generation outages.[2] Moreover, these approaches, while considering the megawatt imbalances, do not address essential characteristics such as ramp (megawatts per minute) and ramp duration uncertainties (minutes), required by the generators participating in the balancing process.

---

[2] An exception is the comprehensive tool developed by Red Eléctrica de España (REE), the Spanish transmission system operator [22].

The methodology developed at PNNL addresses the uncertainty problem comprehensively by including all types of uncertainties (such as load and variable generation) and all aspects of uncertainty including the ramping requirements. The main objective is to provide rapid (every 5 min) look-ahead (up to 5–8 h ahead or day-ahead) assessment of the resulting uncertainty ranges for the balancing effort in terms of the required capacity, ramping capability, and ramp duration. The software production tool has been developed and installed in the CAISO control room for testing and benchmarking.

4.3.1 Methodology

4.3.1.1 Forecast Uncertainties

Uncertainties associated with wind, solar, and load variability, as well as unexpected generation outages, affect load-following needs as well as regulating and contingency reserve requirements. Load forecast uncertainty is one of the most influential factors affecting the resulting uncertainty. System load is normally more significant than wind and solar generation; thus even if the load forecast is more accurate than wind and solar generation forecasts (in terms of percentage errors), the megawatt values of forecast errors can be quite comparable.

Wind generation can be considered as a negative load so that the concept of "net load" is used, which is defined as: net load = total electrical load – total wind generation output + interchange schedule. The resulting forecast error is the difference between the actual and forecasted net load.

4.3.1.2 Assessment of Balancing Capacity Uncertainty

An approach based on time-varying empirical PDF is used to determine combined uncertainty ranges of wind, solar, and load forecast errors. Wind, solar, and load forecast errors are summed for each dispatch interval in the past within a sliding window of a restricted size. The window size is selected to collect sufficient statistical information regarding the CDF of forecast errors. Statistical information is accumulated separately for each forecast horizon, such as hour-ahead forecast and 2-h-ahead forecast. Based on the collected statistics, the approach evaluates percentile intervals (also called confidence intervals or uncertainty ranges) for each forecast horizon. The uncertainty ranges define the interval within which an analyzed parameter is expected to have a specified level of confidence. To determine the uncertainty ranges, it is necessary to find two levels of the inverse CDF function corresponding to the desired percentiles on both ends of the distribution curve. If the CDF is a continuous function, then the inverse CDF function, $CDF-1(p), p \in [0, 1]$, gives a unique real number x such that $CDF(x) = p$. The purpose of the analysis is to identify the net load forecast error range (x_1, x_2) for a given level of confidence p:

$$CDF(x_2) - CDF(x_1) = p(x_1 \leq X \leq x_2) = \int_{x_1}^{x_2} PDF(x)\,dx \qquad (4.9)$$

4.3.1.3 Assessment of Ramping Uncertainties

The assessment of ramping requirements is very important for successful integration of large-scale wind and solar resources into a power system. Sudden variable generation ramps require additional fast responding generation units to be available online.

The required ramping capability needed to follow the net load curve can be derived from the regulation and load-following curves described in Makarov et al. [23]. The regulation capacity and its ramping requirements are inherently related. Insufficient ramping capability could cause additional capacity requirements. A multivariate statistical analysis [24] is applied in this chapter to provide a concurrent consideration on the capacity, ramping, and ramp duration requirements of the regulation and load following. Because of space limitation, a detailed description of the "swinging door" algorithm and the multivariable statistical analysis method are not presented in this chapter. Reference [23] provides a detailed description of the methods.

4.3.1.4 "Flying Brick" Method

Based on the multivariate statistical analysis, a method called the "flying brick" method is proposed to analyze time-varying extreme requirements (the worst cases within the uncertainty range) of the look-ahead generation capacity, ramping capability, and ramp duration [24]. The worst combinations of these parameters are found at the vertices of the flying brick. The objective is to include the capacity, ramp rate, and ramp duration requirements simultaneously and directly into the generation scheduling and dispatch processes.

Figure 4.8 illustrates the idea of the flying brick method. Three uncertainty ranges (i.e., capacity, ramp rate, and ramp duration requirements) are represented as a three-dimensional probability box, that is, the flying brick. Two curves are shown in the figure. The first curve shows the generation requirements that meet the expected net load. The second curve refers to actual net load, which can deviate from its expected values. Suppose t_0 is the current moment. At this point, the multivariable statistical analysis is applied to forecast errors for different look-ahead intervals. The

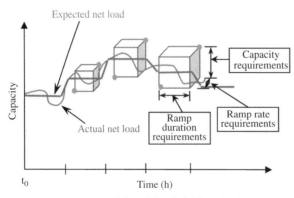

Figure 4.8   The idea of the "flying brick" method.

worst combinations of the three requirements shown by the vertices of the probability box provide generation characteristics needed to meet system requirements with a certain level of confidence. For each subsequent time interval, the probability box is built based on three-dimensional CDFs.

4.3.2 Simulation Results

Using the prototype tool, case studies have been run to test the uncertainty assessment approach and to demonstrate the capabilities of the tool. CAISO's actual data were used in the simulation and tool development. Figure 4.9 presents an example of the snapshot of the real-time capacity requirements for the next operating hour. The light and dark gray uncertainty ranges represent the evaluated capacity ranges for the 90% and 95% confidence levels. These ranges are built around the real-time generation schedule (economic dispatch) curve. The available balancing reserve, which is calculated from the margin of online generators is also shown in Figure 4.9. The system would have adequate balancing reserve with a specified confidence level if the available balancing reserve covers the entire net load uncertainty range. Otherwise, deficiency of balancing reserve occurs. This means that there is certain probability that the online generation will not be able to follow the net load requirement.

The deficiency of the balancing reserve, shown as vertical bars, provides guidance to operators for committing or de-committing additional generators to achieve the desired confidence level for the balancing reserve.

It has been found that the tool is capable of predicting intra-hour deficiency in generation capability. This deficiency of balancing resources can cause price spikes in real-time markets. One can see in Figure 4.9 that more than a 300-MW deficiency of the system generation capability is predicted in the 10–25 min look-ahead period (red error bars). Price spikes occurred at CAISO market at the same time when generation capability deficiency was predicted by the PNNL tool.

Figure 4.10 presents the real-time ramping requirements prediction display. This evaluation is based on statistical analysis of real-time forecast uncertainty. The length of the vertical bars indicates the ramp rate requirements (megawatts per minute). The width of the bars indicates the ramp duration requirements (minutes). This information can be used to determine whether the system would be able to meet the ramp rate requirements. From this figure, operators can compare the ramping requirements against the available system ramping capability to see if the system has sufficient resources to meet the ramp rate requirements. Insufficient ramping capability is indicated using vertical error bars (Figure 4.10).

4.4 QUANTIFYING THE REGULATION SERVICE REQUIREMENTS AT ERCOT

Wind generation adds value to the electric power industry by diversifying generation portfolios, reducing fuel costs, hedging against fluctuating gas prices, and

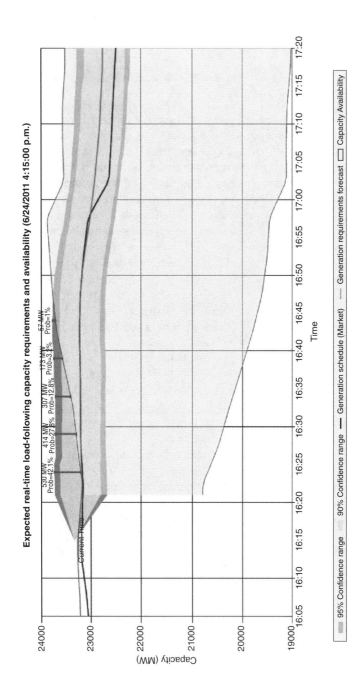

Figure 4.9 Real-time capacity requirement screen.

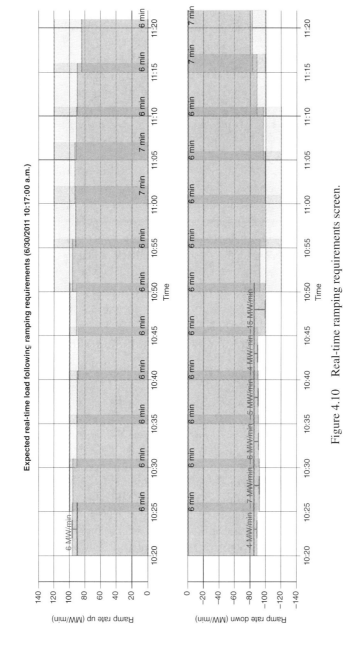

Figure 4.10 Real-time ramping requirements screen.

providing a sustainable source of clean energy. The Electric Reliability Council of Texas (ERCOT) BA leads the United States with over 12 GW of installed wind capacity and nearly 66 MW of peak demand. The ERCOT system is predicted to have over 16 GW installed wind capacity by 2016. The increased variability and uncertainty of wind poses additional operational challenges, such as increased need for reserves. Reserves can be defined as additional capacity and responsive load, either online or off-line, above that needed to meet the electricity demand in order to maintain reliability if actual load or generation differs from what is expected or to account for unexpected changes such as a sudden loss of generation or transmission. The types and characteristics of reserves vary in response speed, whether they are online or off-line, and duration time. ERCOT uses three categories of ancillary services [25].

1. "Regulation up" and "regulation down" are used to control power output of resources in response to small changes in system frequency in order to maintain the system frequency within predetermined limits.
2. Responsive reserve is used to restore ERCOT system frequency in the event that a major event occurs on the grid and significantly affects system frequency. The minimum requirement for this service is 2800 MW and the resources providing responsive reserve must be deployed within 10 min when called.
3. Nonspinning reserve service consists of off-line or reserved capacity, or load resources (interruptible loads), which are capable of deploying within 30 min for at least 1 h. This service intends to cover the contingency of losing resources or under-forecasting load or during extreme temperatures.

These ancillary services are critical to maintaining a balance between generation and load demand, thereby resulting in a satisfactory frequency profile. There is a serious concern that increasing wind penetration in the ERCOT system will have a profound effect on ancillary services. As more wind generation is added to the grid, the operation conditions could vary dramatically from one hour to another. It becomes more efficient to procure the ancillary service when close to the real-time operation conditions because this could help to reduce, if not completely eliminate, the uncertainty in the forecast. On the other hand, load-serving entities tend to know the ancillary service requirement prior to the real time so that they have adequate time to develop hedging strategies against these uncertainties. Therefore, the ancillary service requirement is calculated monthly at ERCOT. To account for the effect of increasing wind generation on the regulation service requirement, a statistics approach was also developed which is detailed subsequently.

4.4.1 Regulation Service (RGS) Requirement

ERCOT has developed a procedure for determination of the base requirement for regulation service. The base requirement is calculated as follows:

Calculate the 98.8 percentile of the 5-min net load (load and wind) changes during the 30 days prior to the time of the study and for the same month of the previous year by hour. Also, calculate the 98.8 percentile of the up and down regulation service deployed during the 30 days prior to the time of study and for the same month of the previous year by hour. These results will be used to calculate the amount of regulation service required by hour to provide an adequate supply of regulation service capability 98.8% of the time.

ERCOT will calculate the increased amount of wind penetration each month and utilize a look-up table to consider the effect of historical wind data on the computation of regulation service requirements. The look-up table indicates additional megawatts to add to the regulation requirements per 1000 MW of increase in wind generation.

If it is determined that during the course of the 30 days prior to the time of the study that the ERCOT average CPS1 score was less than 100%, additional regulation up and down will be procured for hours in which the CPS1 score was less than 100%. Each month ERCOT will perform a back-cast of last month's actual exhaustion rate. If the exhaustion rate exceeded 1.2% in any given hour, ERCOT will determine the amount of increase necessary to achieve an exhaustion rate of 1.2% for that hour.

The procedures to calculate base requirement for regulation services are shown in Figure 4.11.

The amount of monthly regulation-up services requirement calculated for ERCOT is shown in Figure 4.12. The figure shows that there is a clear seasonal and

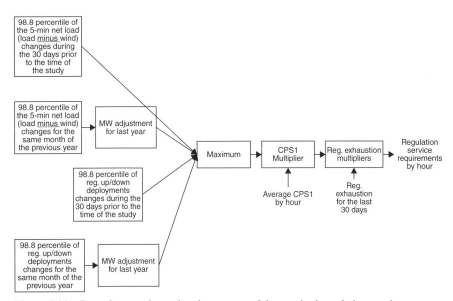

Figure 4.11   Procedures to determine the amount of the required regulation services.

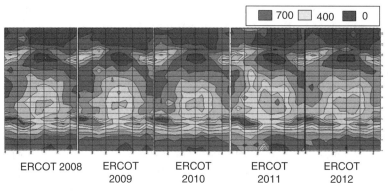

Figure 4.12   Ancillary services requirements in ERCOT.

diurnal pattern in regulation from year to year. The vertical axis lists the months of the year from January to December, and the horizontal axis lists the hours-of-day from H0000 to H2400. Certain patterns can be identified from Figure 4.12.

Summer peak: Increased stress on the power system that occurs in the summer months during mid-day is well documented and it can be observed in the middle of each year (July–August, approximately 8:00–11:00). As weather temperatures climb toward summer highs, consumers increase use of air conditioners, reduce the generation capacity margin, and increase the need for regulation.

Winter peak: Another important time period is the winter peak, or the increased stress that occurs in the winter months in the early evening. This increase in regulation is due to the phenomenon of consumers turning on heaters and other electronics upon returning home from work in winter.

Morning peak: Peaks in regulation that occur in the morning throughout the year except summer have even larger requirements than the summer peak. These increased requirements occur due to variability in generation start-up in the early morning and the initial execution of the new day-ahead schedule.

These seasonal and diurnal trends demonstrate a temporal map of when ancillary services are most needed as a result of the variability in net load. Although the requirements of the power system at midday in summer are well documented, there are other interesting temporal periods that present challenges and merit attention, such as the winter peak and morning peak. This map represents a "snapshot" visualization of both the diurnal and seasonal trends for regulation-up reserve requirements.

4.4.2 Effect of Wind Generation over Regulation Service (RGS) Requirement

A study performed by General Electric (GE) on the ERCOT system used time-series wind and load modeling and statistical analysis to assess the amount of required ancillary services with increasing installed wind capacity [26]. In this study, minute-by-minute time-series wind and load data were used and a statistical analysis was employed to assess how much additional ancillary services will be needed due to increasing installed wind capacity. Both wind and load are time-varying, difficult to perfectly predict, and mutually dependent on weather conditions. This mutual dependence indicates that the variability of both wind and load drives the need for RGS. This means that RGS requirement must be analyzed from the combined data, which is net-load. The net-load defined as "the aggregate customer load demand minus the aggregate wind generation output" is the stochastically varying input that the power system operator must use to dispatch generation and maintain system frequency. By studying the variability of net-load, the chronological factor between the load and wind is preserved. In contrast to other methods where the variances of load and wind are arithmetically added, this approach is more accurate by considering the interdependence between the load and wind.

Since RGS is deployed to compensate for short time-scale fluctuations, the regulation requirements were calculated as a percentage of the historical minute-by-minute deviation of the net-load from the SCED base point. The implicit assumption made here is that given that the SCED is executed every 5 min, adequate ramping capacity from conventional resources is available to allow the dispatch points to be followed very closely. The stochastic, relatively fast minute-by-minute variation of net-load needs to be cancelled out by RGS. In order to isolate the effects of typical increases in regulation due to wind installation (as opposed to wind ramps due to extreme weather events), the uncertainty in the hourly short-term load/wind forecast is omitted, and a persistent net-load forecast delivered 5 min prior to the real time is used to approximate the basic dispatch point. Then, a statistical analysis is performed on the regulation requirements for each wind scenario: the 98th percentile of maximum regulation over 5 min is calculated for each hour-of-day in the year. As a result, this study qualitatively assesses the impact of wind integration on the levels of required ancillary services by looking at the difference between these scenarios.

The main result of this study was that the effects of adding wind capacity result in a nearly perfectly linear increase in regulation requirements. This creates a simple model for calculating the effects of increasing wind penetration on regulation reserves; the incremental increase of regulation requirements due to wind is expressed in wind-impact coefficiencies, which give the incremental change in regulation services per 1000 MW installed wind capacity. The ultimate change resulting from the additional wind capacity is the wind-impact coefficiencies multiplied by the additional wind capacity.

The coefficiencies for impact on regulation-up and regulation-down service caused by 1000 MW additional wind generation can be visualized using a contour map (Figure 4.13). The vertical axis lists the months of the year from January to December, and the horizontal axis lists the hours-of-day from 0 to 24.

Each BA shall operate such that, on a rolling 12-month basis, the average of the clock-minute averages of the BA-ACE divided by $10B$ (B is the clock-minute average of the BA area's frequency bias) times the corresponding clock-minute averages of the interconnection's frequency error is less than a specific limit (NERC BAL-001-0.1a Standard). The score of CPS1 shown in Figure 4.14 indicates good quantity and quality of regulation resources procured since 2012.

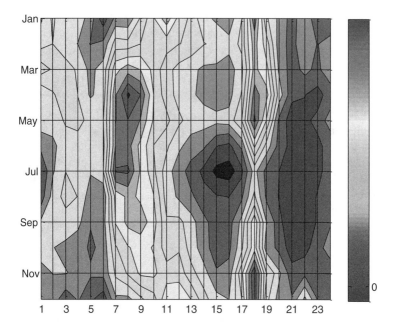

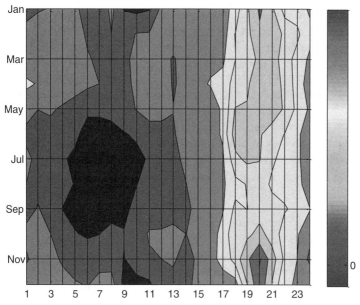

Figure 4.13   Coefficiencies for impact on regulation service caused by 1000 MW additional wind generation (up: regulation up, and bottom: regulation down).

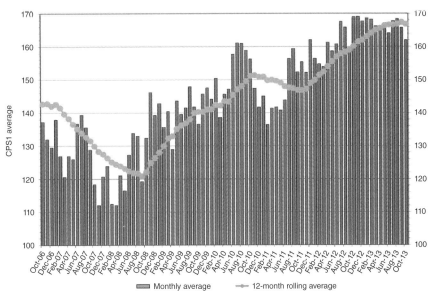

Figure 4.14   Frequency response performance (CPS1) at ERCOT.

4.5 CONCLUSIONS

A power system with large-scale renewable resources like wind/solar power creates significant challenges to the system operation because of intermittency and uncertainties associated with variable generations. It is important to quantify these uncertainties and then incorporate this information into decision-making processes and power system operations, for better reliability and efficiency. This chapter presents three innovations to evaluate the flexibilities needed in the presence of variable generation and how to allocate them accordingly.

A methodology of predicting the impact of wind and solar generation, uninstructed unit deviations, and load uncertainties on the next operating day regulation requirements has been developed. The methodology takes into account CPS and BAAL performance standards. Simulations have shown that BA's savings on the regulation procurement could potentially reach up to between 10% and 30%.

A methodology and software tool for evaluating the wind and solar generation load uncertainties, as well as unexpected generation outages, has been developed. As a result, the uncertainty ranges for the required generation performance envelope can be evaluated for a look-ahead period. The generation performance envelope includes uncertainty ranges for the required balancing capacity, ramping capability, and ramp duration capability. The tool enables prediction of system balancing resources deficiency.

A method has been proposed and evaluated at ERCOT to determine the monthly regulation service requirement, which considers the impact of increasing wind generation. This statistically based approach tends to cover the 98th percentile of the minute-by-minute variation of net-load. While the amount of regulation service requirement is calculated in a static way, this offers advantages in providing a balance between the efficiency and reliability since it allows market participants to hedge against the price volatility ahead of time.

APPENDIX 4.A NOMENCLATURE

NI_A	Actual net interchange (MW)
NI_S	Scheduled net interchange (MW)
B_i	Frequency bias for the balancing authority (BA) (MW/0.1 Hz)
F_A	Actual frequency (Hz)
F_S	Scheduled frequency (normally 60 Hz)
T_{0b}	Bilateral payback for inadvertent interchange (MW)
I_{ME}	Metering error correction (MW)
B_S	WECC system frequency bias (MW/0.1 Hz)
Y	B_i/B_S
$II_{primary}^{on/off\ peak}$	BA's accumulated primary inadvertent interchange (MWh)
H	Number of hours used to pay back inadvertent interchange energy
G_A	Actual conventional generation (without wind and solar)
W_A	Actual wind generation
S_A	Actual solar generation
L_A	Actual load
G_S	Generation dispatch (schedule)
W_S	Wind generation forecast (schedule)
S_S	Solar generation forecast (schedule)
L_S	Load forecast (schedule)
ΔG	Uninstructed deviation of conventional generation units
ΔW	Wind generation forecast error
ΔS	Solar generation forecast error
ΔL	Load forecast error
ACE^*	ACE that would be observed in the case of no regulation in the BA control area
REG_A	Actual regulation applied to the system
ACE_{WECC}	BA-ACE
REG_t^*	Regulation needed to have a zero ACE

$BAAL_{maxt}/BAAL_{mint}$ Limits at time t

ACE_t^* BA's ACE in the case that the system does not have regulation at time t

L_{10} A special limit

REFERENCES

[1] S. Venkataraman, G. Jordan, R. Piwko, L. Freeman, C. Loutan, G. Rosenblum, M. Roth-leder, J. Xie, H. Zhou, and M. Kuo, "Integration of Renewable Resources: Operational Requirements and Generation Fleet Capability at 20% RPS," California ISO, Folsom, CA, August 31, 2010. Available at https://www.caiso.com/Documents/Integration-RenewableResources-OperationalRequirementsandGenerationFleetCapabilityAt20PercRPS.pdf

[2] D. A. Halamay, T. K. A. Brekken, A. Simmons, and S. McArthur, "Reserve requirement impacts of large-scale integration of wind, solar, and ocean wave power generation," *IEEE Transactions on Sustainable Energy*, vol. 2, pp. 321–328, 2011.

[3] R. Masiello, K. Vu, L. Deng, A. Abrams, K. Corfee, and J. Harrison, "Research Evaluation of Wind Generation, Solar Generation, and Storage Impact on the California Grid," Prepared for California Energy Commission Public Interest Energy Research Program, KEMA, June 2010.

[4] A. Mills, M. Ahlstrom, M. Brower, Brower, A. Ellis, R. George, T. Hoff, B. Kroposki, C. Lenox, N. Miller, J. Stein, and Y. Wan, *Understanding Variability and Uncertainty of Photovoltaics for Integration with the Electric Power System*, Ernest Orlando Lawrence Berkeley National Laboratory, Berkeley, CA, December 2009.

[5] D. Lew, M. Milligan, G. Jordan, L. Freeman, N. Miller, K. Clark, and R. Piwko, *How Do Wind and Solar Power Affect Grid Operations: The Western Wind and Solar Integration Study*, National Renewable Energy Laboratory, Golden, CO, September 2009. Available at http://www.nrel.gov/docs/fy09osti/46517.pdf

[6] "Western Wind and Solar Integration Study," Prepared for NREL by GE Energy, May 2010. Available at http://www.nrel.gov/docs/fy10osti/47434.pdf

[7] Y. V. Makarov, C. Loutan, J. Ma, and P. De Mello, "Operational impacts of wind generation on California power systems," *IEEE Transactions on Power Systems*, vol. 24, pp. 1039–1050, May 2009.

[8] NERC. "Real Power Balancing Control Performance" Standard BAL-001-2. Available at http://www.nerc.com/pa/Stand/Reliability%20Standards/BAL-001-2.pdf

[9] N. Jaleeli and L. Van Slyck, "NERC's new control performance standards," *IEEE Transactions on Power Systems* , vol. 14, pp. 1092–1099, August 1997.

[10] Y. Makarov, P. Etingov, N. Samaan, J. Ma, and C. Loutan, "Predicting Day-Ahead Regulation Requirement for CAISO Balancing Area," *California Energy Commission*, Folsom, CA, 2011.

[11] J. Zack, "An Analysis of the Errors and Uncertainty in Wind Power Production Forecasts," Proceedings of the WINDPOWER Conference and Exhibition 2006, Pittsburgh, June 4–7, 2006.

[12] J. Lerner, M. Grundmeyer, and M. Garvert, "The role of wind forecasting in the successful integration and management of an intermittent energy source," *Energy Central, Wind Power*, vol. 3, no. 8, July 2009.

[13] G. Kariniotakis, "ANEMOS, Leading European Union Research on Wind Power Forecasting," Proceedings of the International Wind Forecast Techniques and Methodologies Workshop, July 24–25, 2008.

[14] Energy & Meteo Systems, "Wind Power Prediction Previento." Available at http://energymeteo.de/de/media/e_m_Broschuere.pdf

[15] P. Pinson, G. Kariniotakis, H. A. Nielsen, T. S. Nielsen, and H. Madsen, "Properties of Quantile and Interval Forecasts of Wind Generation and Their Evaluation," Proceedings of the European Wind Energy Conference & Exhibition, Athens, Greece, February 2–March 2, 2006.

[16] A. Luig, S. Bofinger, and H. G. Beyer, "Analysis of Confidence Intervals for the Prediction of the Regional Wind Power Output," Proceedings of the 2001 European Wind Energy Conference, Copenhagen, Denmark, July 2–6, 2001.

[17] J. Kehler, M. Hu, M. McMullen, and J. Blatchford, "ISO Perspective and Experience with Integrating Wind Power Forecasts into Operations," Proceedings of the IEEE General Meeting, Minneapolis, July 25–29, 2010.

[18] D. Maggio, C. D'Annunzio, S.-H. Huang, and C. Thompson, "Utilization of Forecasts for Wind-Powered Generation Resources in ERCOT Operations," Proceedings of the IEEE PES General Meeting, Minneapolis, July 25–29, 2010. Institute of Electrical and Electronics Engineers, Piscataway, NJ.

[19] E. M. Constantinescu, V. M. Zavala, M. Rocklin, S. Lee, and M. Anitescu, "A computational framework for uncertainty quantification and stochastic optimization in unit commitment with wind power generation," *IEEE Transactions on Power Systems*, vol. 26, no. 1, pp. 431–441, February 2011.

[20] J. Wang, M. Shahidehpour, and Z. Li, "Security-constrained unit commitment with volatile wind power generation," *IEEE Transactions on Power Systems*, vol. 23, no. 3, pp. 1319–1327, August 2008.

[21] B. C. Ummels, M. Gibescu, E. Pelgrum, W. L. Kling, and A. J. Brand, "Impacts of wind power on thermal generation unit commitment and dispatch," *IEEE Transactions on Energy Conversion*, vol. 22, no. 1, pp. 44–51, March 2007.

[22] H. Neumann, "Wind Development, Integration Issues and Solutions—TSO Germany," Presentation at the Northwest Wind Integration Forum Technical Work Group: International Large Scale Wind & Solar Integration Techniques & Operating Practice in Germany, Denmark, Spain, July 29–30, 2010 [online]. Available at http://www.nwcouncil.org/media/10999/WIF_TWG_072910_Neumann_070710.pdf

[23] Y. V. Makarov, C. Loutan, J. Ma, and P. De Mello, "Operational impacts of wind generation on California power systems," *IEEE Transactions on Power Systems*, vol. 24, no. 2, pp. 1039–1050, 2009.

[24] Y. Makarov, P. Etingov, J. Ma, Z. Huang, and K. Subbarao, "Incorporating uncertainty of wind power generation forecast into power system operation, dispatch, and unit commitment procedures," *IEEE Transactions on Sustainable Energy*, vol. 2, pp. 433–442, October 2011.

[25] ERCOT, "ERCOT Methodologies for Determining Ancillary Service Requirements," 2014.

[26] General Electric, "Analysis of Wind Generation Impact on ERCOT Ancillary Services Requirements," March 28, 2008.

STOCHASTIC APPLICATIONS

CHAPTER 5

SECURITY-CONSTRAINED UNIT COMMITMENT WITH UNCERTAINTIES

Lei Wu and Mohammad Shahidehpour

RECENT BLACKOUTS IN THE United States and throughout the world have provided a growing evidence that security continues to be the most important aspect of power system operations. The fundamental tool in calculating the limits for the secure and economic operation of power systems is security-constrained unit commitment (SCUC), which has been widely utilized by independent system operators (ISOs) and regional transmission organizations (RTOs) to clear short-term electricity markets. The SCUC solution provides an hourly generation schedule for supplying the power system demand and meeting the security margin, and an hourly generating reserve schedule for satisfying operating constraints when unexpected changes occur in the power system demand or when the power system faces sudden contingencies.

The rapid deployment of variable and distributed renewable generation and the use of flexible demand response (DR) resources in emerging electric power systems have increased the system variability and presented new challenges to ISOs and RTOs and their operations. Such variabilities together with other uncertain factors in power systems, such as random outages of generation and transmission assets and forecast errors pertaining to hourly loads, availability of renewable energy resources, and market prices for electricity and fossil fuels may result in power supply shortages, excessive frequency fluctuations, and even regional brownouts and blackouts. Hence, the impact of potential uncertainties on power system security is of fundamental importance, especially when multiple uncertain factors are considered simultaneously in the power system operation and control. However, incorporating such uncertainties into the stochastic SCUC model is challenging in particular when we have already considered difficulties in the modeling of large-scale deterministic SCUC problem.

This chapter reviews the state-of-the-art modeling and solution methodologies for a deterministic SCUC and extends the results to the solution of SCUC with uncertainties (including the solution methods pertaining to robust optimization-based

Power Grid Operation in a Market Environment: Economic Efficiency and Risk Mitigation, First Edition.
Edited by Hong Chen.
© 2017 by The Institute of Electrical and Electronics Engineers, Inc. Published 2017 by John Wiley & Sons, Inc.

SCUC, stochastic programming-based SCUC, and chance-constrained optimization-based SCUC), for effectively managing uncertainties and enhancing the economics and security of power system operations. Numerical case studies are included in this chapter, which indicate the effectiveness of the listed models and solution methodologies with substantial practical and computational requirements as we consider the operational security of power systems under uncertainties. The major challenge remains to be the computational performance of the SCUC models in practical and large-scale power systems, which indicates that further research is required on such extremely challenging operation problems for improving the economics and security of electric power systems.

5.1 INTRODUCTION

The integration of distributed and renewable energy resources, use of shale gas as a viable option for offering cleaner fossil fuel, advent of smart grid and grid modernization, and major outages in electric power system operations throughout the world have made it clear that additional actions are needed in the electricity sector for continuing to offer clean, secure, resilient, and affordable energy to its customers.

Since the electricity market restructuring started in the 1990s, a significant portion of the electricity infrastructure has been operated under the market environment. The generation business is rapidly becoming market-driven. Market participants, such as generation companies (GENCOs), transmission companies (TRANSCOs), and distribution companies (DISCOs), use available signals for strategically constructing their generation bids, which include signals on locational marginal prices (LMPs) and transmission congestions [1–3]. The ISO coordinates market participants' bids for satisfying hourly load demand, limited fuel and other resources, environmental constraints, and transmission security requirements.

The restructuring of the electricity industry and the creation of self-interested entities such as GENCOs and TRANSCOs have surfaced many shortcomings of the existing electricity systems in an interconnected power network. At the same time, power system security continues to be the most important aspect of the system operation which cannot be compromised in a market-driven approach. Security refers to the ability of a power system to withstand sudden disturbances [4–6]. In a plain language, security implies that a power system remains intact even after credible outages. The standard market design (SMD) is applied to electricity markets for scheduling a secure and economically viable hourly power generation in the day-ahead operation. One of the key components of SMD is SCUC which utilizes the detailed market information submitted by market participants, including the temporal characteristics of generators and transmission lines, availability of transmission capacities, hourly generation offers, scheduled transactions, and curtailment contracts [7–9]. Considering potential outages, SCUC provides an hourly unit commitment (UC) solution that is physically feasible and financially viable.

Climate changes in general and global warming in particular have required environmental issues to be considered in power system operations. The smart grid calls for the evolution in a twenty first century power grid that connects everyone to abundant, affordable, clean, and reliable supply of electricity anytime, anywhere [10]. The smart grid will integrate advanced transactive techniques including all kinds of generation sources, customer participation, among others, into the electric grid. However, a rapid deployment of variable renewable generations and price-sensitive DR resources presents new challenges to ISOs in determining an hourly generation schedule for balancing demand and managing the security of electricity infrastructure.

This chapter reviews existing models and solution methodologies to effectively handle uncertainties in the SCUC problem for enhancing the power system economics and security, with a focus on the short-term operation (i.e., day ahead). The rest of the chapter is organized as follows. Section 5.2 reviews the deterministic SCUC models and solution methodologies. Section 5.3 discusses major uncertainties in emerging power systems. Section 5.4 introduces models and solution methodologies that could effectively handle uncertainties in the SCUC problem. Section 5.5 presents numerical illustrative results, and the concluding comments are drawn in Section 5.6.

5.2 SCUC

The fundamental concept in the secure and economic operation of electric power systems is SCUC. SCUC refers to the economic scheduling of generating units for serving the hourly load demand while satisfying temporal and operational limits of generation and transmission facilities in contingency-based power systems [9, 11]. In a vertically integrated utility, system operators apply SCUC for minimizing the operating cost while meeting the system load by starting up/shutting down generating units. In restructured power markets, SCUC is utilized by ISOs or RTOs to clear real-time and day-ahead markets, with the objective of maximizing the social welfare based on offers and bids submitted by market participants. The SCUC models and solution methodologies should be considered as a key decision-making component in power system operations, especially for large-scale systems. Such SCUC applications are implemented in US electricity markets including MISO, PJM, NYISO, CAISO, ERCOT, and ISO-NE among others.

In order to implement SCUC, forecasted hourly load and system reserve requirements over a given time period as well as physical characteristics and operation constraints of generating units and transmission lines should also be considered as input. The SCUC solution would provide the hourly generation schedule for supplying the system demand and meeting the security margin. The hourly generating reserve schedule satisfies operating constraints in case of contingencies or unexpected changes in demand.

5.2.1 The SCUC Model

The SCUC problem formulation is described as follows [12]. The objective of the SCUC is to minimize the total operating cost based on generating unit offers and load demand bids. The objective function in equation (5.1) includes incremental cost, no-load cost, and start-up/shutdown cost, where i, t, and k are the indices for units, hours, and segments of cost curve functions. c_{ik} is the incremental cost for segment k of unit i. N_i is the no-load cost of unit i. P_{ikt} is the generation dispatch of unit i at hour t at segment k. I_{it} is the UC decision of unit i at hour t. SU_{it} and SD_{it} are the start-up and shutdown costs of unit i at hour t:

$$\text{Min} \sum_i \sum_t \left[\sum_k c_{ik} P_{ikt} + N_i I_{it} + SU_{it} + SD_{it} \right] \tag{5.1}$$

In general, SCUC constraints include those of base and contingency cases. For the base case, UC constraints include system constraints such as power balance (5.2), and spinning and operating reserves (5.3). w and v are the indices for wind farms and grid-scale battery storages. D_t, R_{St}, and R_{Ot} are the system load demand, and spinning and operating reserve requirements at hour t. P_{it} and P_{wt} are the generation dispatch of thermal unit i and wind farm w at hour t, respectively. OR_{it} and SR_{it} are operating and spinning reserves provided by unit i at hour t. $P_{dc,vt}$ and $P_{c,vt}$ are the power discharged and charged by battery v at hour t. OR_{vt} and SR_{vt} are the operating and spinning reserve provided by battery v at hour t:

$$\sum_i P_{it} + \sum_w P_{wt} + \sum_v (P_{dc,vt} - P_{c,vt}) = D_t \tag{5.2}$$

$$\sum_i SR_{it} + \sum_v SR_{vt} \geq R_{St}$$

$$\sum_i OR_{it} + \sum_v OR_{vt} \geq R_{Ot} \tag{5.3}$$

For the base case, the general UC constraints also include ramping up/down limits as well as start-up/shutdown ramping for individual units in equation (5.4), minimum on/off time constraints in equation (5.5), power generation dispatch limits in equation (5.6), and spinning/operating reserve constraints of individual generators in equation (5.7). Only single-cycle thermal units are modeled in equations (5.4)–(5.7). The formulation of other types of generating units including combined-cycle, and cascaded and pumped storage hydro can be found in References [9, 13]. Equation (5.8) is the generation dispatch limit of wind farms, where $P_{f,wt}$ is the forecasted wind power generation. DR_i and UR_i are the ramping down/up limits of unit i. DP_i and UP_i are the shutdown/start-up ramp limits of unit i. P_i^{min} and P_i^{max} are the minimum/maximum generation capacities of unit i. T_i^{on} and T_i^{off} are the minimum on/off time limits of unit i. X_{i0}^{on} and X_{i0}^{off} are the initial on/off time of unit i. DT_i and UT_i are the initial hours unit i must be off/on due to its minimum off/on time limits. a, b, and m are the indices for buses. P_{ik}^{max} is the capacity of segment k of unit i. MSR_i and QSC_i are the spinning reserve provided in one minute and the quick-start capacity of unit i.

The grid-scale energy storage can handle the peak-hour load, enhance power system security, and utilize the variability of renewable energy. Equations

(5.9)–(5.14) describe battery storage constraints; however, the formulation of other types of energy storage can be similarly included. Constraint (5.9) represents the energy balance in a battery. Equation (5.10) describes the exclusive charge/discharge status, equations (5.11) and (5.12) show charge/discharge capacity limits, and (5.13) represents the energy limits of a battery. Constraint (5.14) limits initial/terminal energy levels to be equal in a battery. E_{vt} is the energy stored in battery v at hour t. $P_{dc,vt}$ and $P_{c,vt}$ are the discharged/charged power in battery v at hour t. η_v is the efficiency of battery v. $I_{dc,vt}$ and $I_{c,vt}$ are charge/discharge indicators in battery v at hour t. $P_{c,v}^{min}$ and $P_{c,v}^{max}$ are lower/upper charge limits of battery v. $P_{dc,v}^{min}$ and $P_{dc,v}^{max}$ are lower/upper discharge limits in battery v at hour t. E_v^{min} and E_v^{max} are the lower/upper energy limits in battery v. E_{v0} and E_{vNT} are the initial/terminal energy levels of battery v.

For the base case, equation (5.15) represents the DC network constraints using the linear sensitivity factor (LSF) method, which represents the sensitivity of line flows to bus power injections [9]. In equation (5.15), LSF_l^m represents the sensitivity of power flow on line l to power injection at bus m, and $(\sum_{i:i\in U(m)} P_{it} + \sum_{w:w\in U(m)} P_{wt} + \sum_{v:v\in U(m)}(P_{dc,vt} - P_{c,vt}) - D_{mt})$ calculates the net real power injection at bus m. AC network constraints can also be incorporated into SCUC [7, 14]. PL_{lt} is the power flow of line l at hour t, and PL_l^{max} is the power flow limit of line l. $U(m)$ is the set of generators/wind farms/batteries located at bus m. D_{mt} is the load of bus m at hour t:

$$
\begin{aligned}
P_{it} - P_{i(t-1)} &\leq \text{UR}_i I_{i(t-1)} + \text{UP}_i\left[I_{it} - I_{i(t-1)}\right] + P_i^{max}[1 - I_{it}] \\
P_{i(t-1)} - P_{it} &\leq \text{DR}_i I_{it} + \text{DP}_i\left[I_{i(t-1)} - I_{it}\right] + P_i^{max}\left[1 - I_{i(t-1)}\right]
\end{aligned}
\tag{5.4}
$$

$$
\sum_{t=1}^{\text{UT}_i}(1 - I_{it}) = 0 \quad \text{where UT}_i = \max\left\{0, \min\left[NT, \left(T_i^{on} - X_{i0}^{on}\right) I_{i0}\right]\right\}
$$

$$
\sum_{\tau=t}^{t+T_i^{on}-1} I_{i\tau} \geq T_i^{on}[I_{it} - I_{i(t-1)}] \quad \forall t = \text{UT}_i + 1, \ldots, NT - T_i^{on} + 1
$$

$$
\sum_{\tau=t}^{NT}\left[I_{i\tau} - \left(I_{it} - I_{i(t-1)}\right)\right] \geq 0 \quad \forall t = NT - T_i^{on} + 2, \ldots, NT
$$

$$
\sum_{t=1}^{\text{DT}_i} I_{it} = 0 \quad \text{where DT}_i = \max\left\{0, \min\left[NT, \left(T_i^{off} - X_{i0}^{off}\right)\left(1 - I_{i0}\right)\right]\right\}
\tag{5.5}
$$

$$
\sum_{\tau=t}^{t+T_i^{off}-1}\left(1 - I_{i\tau}\right) \geq T_i^{off}[I_{i(t-1)} - I_{it}] \quad \forall t = \text{DT}_i + 1, \ldots, NT - T_i^{off} + 1
$$

$$
\sum_{\tau=t}^{NT}\left[1 - I_{i\tau} - \left(I_{i(t-1)} - I_{it}\right)\right] \geq 0 \quad \forall t = NT - T_i^{off} + 2, \ldots, NT
$$

$$P_{it} = \sum_k P_{ikt}$$

$$0 \leq P_{ikt} \leq P_{ik}^{\max} I_{it}$$

$$P_i^{\min} I_{it} \leq P_{it} \leq P_i^{\max} I_{it} \qquad (5.6)$$

$$P_{it} + SR_{it} \leq P_i^{\max} I_{it}$$

$$SR_{it} \leq 10 \, MSR_i I_{it}$$

$$OR_{it} = SR_{it} + (1 - I_{it}) QSC_i \qquad (5.7)$$

$$0 \leq P_{wt} \leq P_{f,wt} \qquad (5.8)$$

$$E_{vt} = E_{v(t-1)} - (P_{dc,vt} - \eta_v P_{c,vt}) \qquad (5.9)$$

$$I_{dc,vt} + I_{c,vt} \leq 1 \qquad (5.10)$$

$$I_{c,vt} P_{c,v}^{\min} \leq P_{c,vt} \leq I_{c,vt} P_{c,v}^{\max} \qquad (5.11)$$

$$I_{dc,vt} P_{dc,v}^{\min} \leq P_{dc,vt} \leq I_{dc,vt} P_{dc,v}^{\max} \qquad (5.12)$$

$$E_v^{\min} \leq E_{vt} \leq E_v^{\max} \qquad (5.13)$$

$$E_{v0} = E_{vNT} \qquad (5.14)$$

$$-PL_l^{\max} \leq PL_{lt} = \sum_m LSF_l^m \left(\sum_{i:i \in U(m)} P_{it} + \sum_{w:w \in U(m)} P_{wt} \right.$$

$$\left. + \sum_{v:v \in U(m)} (P_{dc,vt} - P_{c,vt}) - D_{mt} \right) \leq PL_l^{\max} \qquad (5.15)$$

Considering contingency cases, thermal generating unit redispatch constraints within an allowable range Δ_{it}^c (5.16)–(5.18) as well as those for wind farms (5.19) and batteries (5.20)–(5.23), and transmission network security constraints (5.24) are included in the SCUC problem. All symbols with superscript c correspond to contingency c:

$$\sum_i P_{it}^c + \sum_w P_{wt}^c + \sum_v \left(P_{dc,vt}^c - P_{c,vt}^c \right) = D_t^c \qquad (5.16)$$

$$P_i^{c,\min} I_{it} \leq P_{it}^c \leq P_i^{c,\max} I_{it} \qquad (5.17)$$

$$\left| P_{it}^c - P_{it} \right| \leq \Delta_{it}^c \qquad (5.18)$$

$$0 \leq P_{wt}^c \leq P_{f,wt}^c \qquad (5.19)$$

$$E^c_{vt} - E^c_{v(t-1)} - \left(P^c_{dc,vt} \quad \eta_v P^c_{c,vt} \right) \tag{5.20}$$

$$I_{c,vt} P^{c,\min}_{c,v} \leq P^c_{c,vt} \leq I_{c,vt} P^{c,\max}_{c,v} \tag{5.21}$$

$$I_{dc,vt} P^{c,\min}_{dc,v} \leq P^c_{dc,vt} \leq I_{dc,vt} P^{c,\max}_{dc,v} \tag{5.22}$$

$$E^{c,\min}_v \leq E^c_{vt} \leq E^{c,\max}_v \tag{5.23}$$

$$-\mathrm{PL}^{c,\max}_l \leq \mathrm{PL}^c_{lt} = \sum_m \mathrm{LSF}^{c,m}_l \left(\sum_{i:i\in U(m)} P^c_{it} + \sum_{w:w\in U(m)} P^c_{wt} \right.$$
$$\left. + \sum_{v:v\in U(m)} \left(P^c_{dc,vt} - P^c_{c,vt} \right) - D^c_{mt} \right) \leq \mathrm{PL}^{c,\max}_l \tag{5.24}$$

The SCUC formulation (5.1)–(5.24) is based on a single binary variable to describe the UC status and the corresponding hourly transition of generating units. Alternatively, a three-binary-variable formulation [15], considering UC status and start-up/shutdown indicators, can be explored in the SCUC formulation with uncertainties. The approach might tighten the MIP solution of the SCUC problem, especially when considering minimum up/down time constraints of thermal units, which are longer than 1 h.

5.2.2 SCUC Solution Methods

Mathematically, the SCUC problem stated in (5.1)–(5.24) is a large-scale mixed-integer optimization problem with a large number of binary variables, continuous and discrete control variables, and a series of prevailing equality and inequality constraints. Considering the computational complexity, SCUC is in a class with non-deterministic polynomial-time hard (NP-hard) problems. Therefore, efficient algorithms have to be developed for attaining optimal/near-optimal SCUC solutions. Considering multi-hour SCUC problems with thousands of generating units and transmission lines, we consider a solution shown in Figure 5.1 which decomposes the SCUC problem into a master problem (hourly UC) and subproblems (transmission network security evaluation). The iterative process between the master problem and the base case network evaluation subproblems (Loop A) will continue until the UC solution with base case transmission constraints is converged. Loop B examines contingencies and prescribes pre-contingency (preventive) and post-contingency (corrective) control actions for maintaining the transmission network security. The contingency evaluation subproblems examine the viability of the base case solution when contingencies are considered and possibly modify the base case solution to maintain the transmission network security. If Loop B fails to eliminate violations in case of contingencies, Loop A will be executed again to update the UC solution based on the information derived from Loop B. Such iterative calculations between the master problem and subproblems continue until no further solution violations exist.

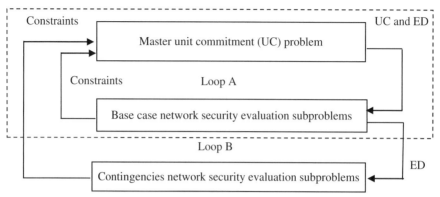

Figure 5.1   Decomposition strategy for solving SCUC.

Various optimization techniques, such as enumeration, priority listing, dynamic programming (DP), Lagrangian relaxation (LR), mixed-integer programming (MIP), and heuristic-based methods (e.g., genetic algorithms, artificial neural networks, expert and fuzzy systems), have been adopted for near-optimal solutions. However, bottlenecks including the enumeration calculation, DP's high dimensionality, and heuristic solution's fine-tuning are considered barriers to practical applications of UC. With the development of more rigorous optimization techniques, LR and MIP are widely applied to day-ahead and real-time generation scheduling problems. The merits of LR and MIP for UC applications are listed below [13, 16].

1. *Modeling*: The LR method can directly model nonlinear cost functions (e.g., quadratic energy consumption costs and highly nonlinear start-up/shutdown costs) and constraints (e.g., fuel and emission constraints). However, the MIP method would require a linearized set or functions. On the other hand, it would be rather easy to add constraints to the MIP model as compared with that in the LR method which would require additional Lagrangian multipliers.

2. *Feasibility and optimality*: The LR solution is theoretically suboptimal due to the non-convexity of the UC problem. In addition, heuristic methods are generally required to tighten the lower bound for a feasible UC solution. As the number of UC constraints grows, the inclusion of a large number of multipliers could render the UC optimization impractical. Mathematically speaking, the MIP solver can guarantee a globally optimal or near-optimal solution, with a gap representing the absolute relative distance between the best solution and relaxed linear programming (LP) solutions. However, it may take longer to find an initial feasible solution of MIP.

3. *Computer resource consumption*: Because the number of DP states and feasible transitions for solving a single UC is proportional to the number of generating units and hours, the required memory and CPU time for the LR solution increase almost linearly with the size of UC problem. In comparison, the required memory and CPU time for the MIP solution are more exponential and

case-dependent. The basic idea for solving the MIP problem is still based on the enumeration method which needs more memory in the branch-and-cut tree method and more CPU time for branching and bounding.

4. Hot-start capability: Considering an iterative SCUC procedure, a feasible initial solution is used to hot-start the MIP solver for reducing the branching and saving the CPU time in a UC problem. However, it is not necessary to flat-start the Lagrangian iteration in the LR application since previous Lagrangian multipliers can accelerate the convergence of the UC solution.

5. Applications: For small-size UC problems, LR and MIP methods have complementary solution merits, although the MIP model may offer a cheaper solution. The LR model offers a good alternative for long-term UC studies, which is due to its higher solution speed, while the MIP method would have to utilize numerical strategies, such as variable time step, rolling horizon, relaxation of integer constraints, and aggregation, for reducing the problem scale.

The base case hourly DC network security evaluation problems(shown in Loop A of Figure 5.1) and those of contingencies (shown in Loop B of Figure 5.1) check possible transmission network violations for the given generation schedule. The transmission system violations can be formed through either the LSF method or the Benders cut, which are added as additional constraints to the master problem. Several features of the two transmission solution methods are summarized below [16, 17].

1. Additional constraints: Linear violation constraints are generated by the LSF method at each period, which may slow down the next UC solution. On the other hand, the Benders cut method offers only one cut (or a predefined number of cuts in the strong Benders cut method) in each period, which generally results in a higher number of iterations than that of the LSF method for the SCUC solution.

2. CPU time: The LP-based subproblem using the Benders cut method takes a longer CPU time than that of the LSF method. On the other hand, the LSF method requires the pre-calculation of the LSF matrix, which may be computationally expensive especially when we consider transmission network contingencies in large-scale power systems.

5.3 UNCERTAINTIES IN EMERGING POWER SYSTEMS

Power system uncertainties include random outages of system components, variability of renewable energy resources (such as natural water inflow, wind speed, and solar radiation), and load and market price forecast inaccuracies. The impact of uncertainty on the operational security of power systems is of fundamental importance when multiple factors are considered in power systems. For instance, when the wind energy integration reaches a critical level, the dependency of power systems on wind energy can inevitably result in additional supply risks associated with the wind energy availability. Furthermore, when demand response reaches a critical market level, the inaccuracy of load forecast could inevitably pose real-time electricity balance risks. Thus, power systems would have to plan for alternate backup generation to ensure the power secure operation in case the real-time status of power systems deviates a lot from the expected forecasted values. This section discusses major uncertain factors in the operation of modern power systems.

5.3.1   Random Outages of System Components

Here, we first discuss random outages of individual power system components as independent events. Let p_i be the availability of the ith component, $q_i = 1 - p_i$ be its unavailability (also referred to as forced outage rate). μ_i and λ_i are the ith component's repair and failure rates. F_i and R_i are the ith component's mean times to failure and repair, respectively. The relationship among these variables is given in equation (5.25):

$$\mu_i = \frac{1}{R_i}, \lambda_i = \frac{1}{F_i}$$

$$p_i = \frac{F_i}{F_i + R_i} = \frac{\mu_i}{\mu_i + \lambda_i}, \quad q_i = 1 - p_i = \frac{R_i}{F_i + R_i} = \frac{\lambda_i}{\mu_i + \lambda_i} \quad (5.25)$$

The probability density functions for operation and failure states are described by exponential distributions as $\lambda_i e^{-\lambda_i t}$ and $\mu_i e^{-\mu_i t}$, respectively. Accordingly, one can calculate the probabilities of a component being available and unavailable as $e^{-\lambda_i t}$ and $e^{-\mu_i t}$, respectively. The state transitions of system components are shown in Figure 5.2

There are several ways to simulate random outages of system components, which are discussed as follows. We use S_{it} to represent the state of component i at time t, where $i = 1, 2, \ldots, N$. $S_{it} = 1$ means the component i is available at time t, otherwise it is on outage. The vector $[S_{1t}, S_{2t}, \ldots, S_{Nt}]$ describes the system state at time t.

a. *State sampling at each time point*

Each component's state is determined by sampling the probability that the component resides in that state, that is, at each time, a uniformly distributed random number U_{it} is generated for each component. Then, U_{it} is compared with the

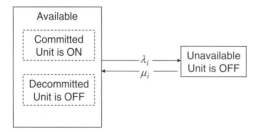

Figure 5.2   State transitions of system component.

forced outage rate q_i. If $U_{it} < q_i$, the component i is unavailable at time t, and $S_{it} = 0$. Otherwise, it is available and $S_{it} = 1$.

b. *State duration sampling*

Each component's state is determined by sampling the duration that the component stays in its current state. The initial state of each component is known and the remaining time in this state is sampled from an exponential distribution with the given failure/repair rate parameters. Then, we continue to generate the successive duration remaining time in failure or repair state alternately, until the time span is ended. For example, if the initial state of a component is available, then the duration of remaining available is sampled from an exponential distribution with a failure rate T_{i1}. Then, the next state is unavailable and the duration for remaining in the unavailable state is sampled from an exponential distribution with a repair rate T_{i2}. This process would continue until $\sum_k T_{ik} \geq \text{TSpan}$. In turn, $S_{it} = 1$ for $t \in T_{i1}$, $S_{it} = 0$ for $t \in T_{i2}$, and so on. The details are shown in Figure 5.3.

c. *State transition sampling*

Similar to state duration sampling, state transition sampling takes into account the history of states. Consider an incremental interval of time dt which is made

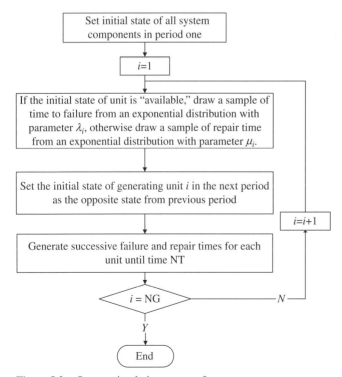

Figure 5.3   Outage simulation process for system components.

sufficiently small so that the probability of two or more state transitions occurring during dt is negligible. We calculate the associated conditional probabilities for component i, that is, the probability of being in the available state at time $(t + dt)$ is the sum of probability of being available at time t and not on outage during time dt, and that of being unavailable at time t, repaired during time period dt, and available at time $(t + dt)$. The conditional probabilities for state transition are shown in equation (5.26):

$$p(\varphi_{t+dt} = 1 \mid \varphi_t = 1) = p_i + q_i\, e^{-(\mu_i+\lambda_i)dt}$$
$$p(\varphi_{t+dt} = 1 \mid \varphi_t = 0) = p_i - p_i\, e^{-(\mu_i+\lambda_i)dt}$$
$$p(\varphi_{t+dt} = 0 \mid \varphi_t = 1) = q_i - q_i\, e^{-(\mu_i+\lambda_i)dt} \tag{5.26}$$
$$p(\varphi_{t+dt} = 0 \mid \varphi_t = 0) = q_i + p_i\, e^{-(\mu_i+\lambda_i)dt}$$

Thus, different from the state duration sampling method, the duration of each state is sampled using the conditional probabilities with the given failure/repair rate parameters. This process continues to generate successive failure and repair times until the time span is ended.

State sampling is simple which only needs state probabilities, but it has the disadvantage of failing to represent the dependency on previous states. The disadvantages of state duration sampling and state transition sampling are the corresponding computation burden and parameter requirements. A hybrid model of state sampling and state duration sampling techniques can be used. Accordingly, for each component i, a uniformly distributed random number U_i is generated. If $U_i < q_i$, the component i is unavailable and the duration of staying unavailable is sampled from an exponential distribution with repair rate. Otherwise, the component is available and the duration of remaining available is sampled from an exponential distribution with failure rate. This process continues successively until the time span is ended.

It must be pointed out that other types of outages, such as a simultaneous failure of two overhead lines on the same tower, multiple component outages caused by an adverse weather condition or a major storm disaster, and cascading outages induced by the lack of protection coordination, can also be simulated through similar procedures. The failure rates of most components will be affected by the weather since they are exposed to varying weather conditions. The common cause outages here mean that outages are not independent and one single event causes multiple components to fail, such as transmission lines connected to the same tower are tripped simultaneously due to a malfunction. Both weather-relatedand common cause outages can increase component failure rates.

a. *Common cause outages*

Figure 5.4 presents common cause outages by introducing common failure and repair rates λ_{ij} and μ_{ij} for components i and j. By combining the simulation of random outages for individual components i or j, and the common cause outage for i plus j, one can simulate the system status by considering individual outages and common cause outages simultaneously.

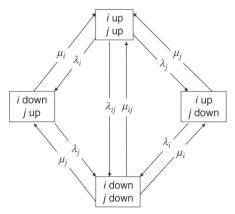

Figure 5.4   State transition diagram with common cause outages.

The analyses can be easily extended to the third or higher order events provided appropriate transitions among states can be identified. Assume the failure rate λ' and the repair rate μ' correspond to both components i and j on outage. The corresponding probability is $\frac{\lambda'}{\lambda'+\mu'} = \frac{\lambda_i}{\lambda_i+\mu_i} \frac{\lambda_j}{\lambda_j+\mu_j}$. Also, $\mu' = \mu_i + \mu_j$, so $\lambda' = \frac{\lambda_i \cdot \lambda_j \cdot (1/\mu_i + 1/\mu_j)}{1 + \lambda_i/\mu_i + \lambda_j/\mu_j}$. Thus, the failure rate in Figure 5.4 is $\lambda = \lambda' + \lambda_{12} = \frac{\lambda_i \lambda_j (1/\mu_i + 1/\mu_j)}{1+\lambda_i/\mu_i+\lambda_j/\mu_j} + \lambda_{12}$.

b. *Weather-related outages*

More parameters are introduced to include the weather effect into the consideration. Let T be the total time duration in a study horizon. N and A represent the total normal and adverse weather in T, that is, $N + A = T$. $\lambda_{i,N}$ and $\lambda_{i,A}$ are the failure rates of component i under normal and adverse weather conditions. Thus, an average failure rate for component i can be expressed by equation (5.27),

$$\lambda_i = \frac{N}{N + A} \lambda_{i,N} + \frac{A}{N + A} \lambda_{i,A} \tag{5.27}$$

5.3.2 System Load Forecast Inaccuracy

Traditional short-term load forecast methods have a reasonable performance with errors of 1–2% [9]. However, short-term load forecasting with the integration of price-sensitive DR is far from reaching maturity. DR integration introduces price-related demand variations which makes the load profile variations price sensitive and stochastic. Under this circumstance, the impact of electricity prices on system loads becomes more visible, which was not the case in a traditionally price-insensitive environment. Thus, price-sensitive load forecasting may need additional price-related information.

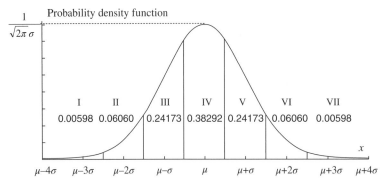

Figure 5.5   Seven-segment approximation of load forecast inaccuracy.

In this regard, the state-of-the-art price forecasting errors are within 5–20% [9], which may deteriorate the accuracy of price-sensitive load forecasts.

Load forecast uncertainties can be represented adequately by a normal distribution in which the mean is the forecasted load. Additional forecasting provisions are considered as follows:

Approach 1. One way is to divide the normal distribution into a discrete number of intervals and the load representing the mid-point of each interval is assigned the probability for that interval. To simplify the calculation, a seven-step distribution $(0, \pm\sigma, \pm2\sigma, \pm3\sigma,)$ is often used where σ, the standard deviation, is 2% of the expected load as shown in Figure 5.5. Such a model will encompass more than 99% of load uncertainty. Table 5.1 shows loads and their corresponding probabilities with the standard deviation of 2%, where L_t^F is the forecasted load at time t.

Approach 2. Since areas of $(-\infty, -3.5\sigma)$ and $(3.5\sigma, +\infty)$ are excluded, the total probability covered by $[-3.5\sigma, 3.5\sigma]$ is less than 1. One way of fixing the issue is to use the truncated normal distribution (5.28). Hence, the total probability would be equal to 1. The probabilities by truncated normal distribution are shown in Table 5.2 as compared with Table 5.1:

TABLE 5.1   Load value and probability pairs for load forecast inaccuracy

Zone	Load value	Probability	Cumulative probability
1	$0.94L_t^F$	0.00598	0.00598
2	$0.96L_t^F$	0.06060	0.06658
3	$0.98L_t^F$	0.24173	0.30831
4	L_t^F	0.38292	0.69123
5	$1.02L_t^F$	0.24173	0.93296
6	$1.04L_t^F$	0.06060	0.99356
7	$1.06L_t^F$	0.00598	0.99954

TABLE 5.2 Load value and probability pairs by the truncated normal

Zone	Load value	Probability	Cumulative probability
1	$0.94L_t^F$	0.00598	0.00598
2	$0.96L_t^F$	0.06063	0.06661
3	$0.98L_t^F$	0.24184	0.30845
4	L_t^F	0.3831	0.69155
5	$1.02L_t^F$	0.24184	0.93339
6	$1.04L_t^F$	0.06063	0.99402
7	$1.06L_t^F$	0.00598	1

$$f(x) = \frac{1}{\sqrt{2\pi}\sigma} e^{-\frac{1}{2\sigma^2}(x-\mu)^2} \bigg/ \int_{-3.5\sigma}^{3.5\sigma} \left[\frac{1}{\sqrt{2\pi}\,\sigma} e^{-\frac{1}{2\sigma^2}(t-\mu)^2} \right] dt \qquad (5.28)$$

Approach 3. The other way is to use the censored method shown in equation (5.29). Accordingly, the total probability is also equal to 1. The corresponding probabilities are shown in Table 5.3 as compared with Table 5.1:

$$x = \begin{cases} -3.5\sigma & x' < -3.5\sigma \\ x' & -3.5\sigma \leq x' \leq 3.5\sigma \\ 3.5\sigma & x' > -3.5\sigma \end{cases} \qquad (5.29)$$

5.3.3 Renewable Energy Resources

Renewable energy resources present major uncertainties in power system operations. In practice, variability, unpredictability, and intermittency are used interchangeably for describing renewable energy situations. Here, we use natural water inflow and wind energy output as two examples to show how the uncertainty of renewable energy resources would be simulated.

TABLE 5.3 Load value and probability pairs by the censored method

Zone	Load value	Probability	Cumulative probability
1	$0.94L_t^F$	0.00621	0.00621
2	$0.96L_t^F$	0.06060	0.06681
3	$0.98L_t^F$	0.24173	0.30854
4	L_t^F	0.38292	0.69146
5	$1.02L_t^F$	0.24173	0.93319
6	$1.04L_t^F$	0.06060	0.99379
7	$1.06L_t^F$	0.00621	1

5.3.3.1 Natural Water Inflow

It is assumed that the water inflow to a reservoir follows a discrete Markov chain which is independent of inflows to other reservoirs. Two stochastic and distributed models, log-normal and Pearson type-3, are used to describe river inflows. The log-normal distribution model is presented to simulate the average river water inflow in seasonal, monthly, or weekly periods. In order to simulate the hourly water inflow profile, we consider $Z_{j,k,t} = Z_{j,k} + \zeta_{j,k,t}$ where $\zeta_{j,k,t}$ is a normal-distributed random number with zero mean and 10% deviation $N(0, 0.1)$ and $Z_{j,k}$ is calculated via equation (5.30):

$$Z_{j,k} = \rho_{j,k-1} Z_{j,k-1} + \left(1 - \rho_{j,k-1}^2\right)^{1/2} \varepsilon_{j,k} \tag{5.30}$$

$$\rho_{j,k-1} = \frac{\sum_{t=1}^{T}(w_{j,k,t} - \overline{w}_{j,k})(w_{j,k-1,t} - \overline{w}_{j,k-1})}{\sqrt{\sum_{t=1}^{T}(w_{j,k,t} - \overline{w}_{j,k})^2 \sum_{t=1}^{T}(w_{j,k-1,t} - \overline{w}_{j,k-1})^2}} \tag{5.31}$$

In (5.30), $Z_{j,k}$ is the standard normal random variable $N(0,1)$. $\rho_{j,k-1}$ is the time serial correlation coefficient for inflows in periods $k - 1$ and k, which can be calculated via equation (5.31), where $\overline{w}_{j,k}$ is the average river water inflow for reservoir j in period k (each period covers T time span). $\varepsilon_{j,k}$ is independent identically random variables following the distribution of $N(0,1)$. Also, $Z_{j,k} = (w_{j,k} - \mu_{j,k})/\sigma_{j,k}$ where $w_{j,k} = \ln(y_{j,k})$. Here, $y_{j,k}$ is the river water inflow to reservoir j in period k; $\mu_{j,k}$ and $\sigma_{j,k}$ are the mean and standard deviation of $w_{j,k}$, respectively.

5.3.3.2 Wind Power Generation

Several studies have investigated the effect of wind power generation variability on the power system security. Some models have relied on the 3σ rule, which assumes that the wind power generation follows the normal distribution and deviates from the expected value by no more than three times of the standard deviation. Wind speed also presents seasonal variations as well which makes it non-stationery in nature, that is, its mean and variance change over time. It is shown that the wind power generation at different granularities and locations would approximately follow certain distributions such as beta, Cauchy, or Weibull.

Time series models have also been used almost exclusively for the simulation of short-term wind power generation [18, 19]. Autoregressive moving average (ARMA) models and their generalizations are used for modeling the temporal dependence structure of wind speed, and multivariate probability distribution functions have been used for modeling the spatial distribution dependence of wind speed variations. Since the autocorrelation factor (ACF) and the partial autocorrelation factor (PACF) of wind speed time series decrease dramatically as the time lag increases, the hourly wind speed forecast error is usually represented by a lower order ARMA (1,1). The ARMA constants are acquired by minimizing the root mean square error (RMSE) between the simulated ARMA time series and the measured wind speed data. Alternatively,

Markov chain is used to simulate wind data, in which the wind speed is classified into several ranges according to its mean value at each category, and the probability transition matrix is applied to define transitions from one wind speed category to others.

Other studies have focused on characterizing wind variability in the frequency domain [20–22]. However, the wind power spectrum may lose all its temporal information because it merely provides the information on amplitudes but not the associated phases, which makes it difficult to derive useful conclusions on the wind speed variations in the time domain. In addition, wind speed variability metrics relying on the power spectral density lack a clear connection to power system operations.

5.3.4 Interdependence of Natural Gas and Electricity Infrastructures

Natural gas and renewable energy are the two most vital energy resources in the electric power industry's transition to an environmental-friendly operation. The use of natural gas and renewable energy in electric power sector has grown significantly in recent years. The attractive utilization of shale gas has introduced the lowest natural gas prices in a decade, which may further expand investments on natural gas generating plants in electric power systems. According to the Clean Power Plan 2015 released by the White House and the U.S. Environmental Protection Agency [23], the power sector is targeted to reduce 32% from its 2005 CO_2 emission level by 2030, and 90 GW coal-fired fleet, almost 30% of total coal-fired capacity, and 8.5% of total generation capacity, will be retired by 2040. This large retired capacity will be filled by new gas-fired generators and renewable generation. Indeed, natural gas has shown a sharp growth in the generation mix, from 18% in 2005 to 27% in 2014, while coal is decreased from 50% to 39%.

Natural gas and renewable energy operations appear to be complementary in many respects concerning variability and environmental impacts at load centers. Hence, the coordination between natural gas and renewable energy would need to be reinforced in integrated resource planning. Specifically, the coordination of constrained electricity and natural gas infrastructures could firm the variability of wind energy in electric power systems. The abundance of sustainable wind energy could substitute natural gas-fired generating units which are constrained by fuel availability and emission. Also, the flexibility and quick ramping capability of natural gas units could firm the variability of wind energy. Accordingly,

- Natural gas has experienced significant price variations which would directly affect the cost of commitment and generation of thermal generating units [24]. While the renewable generation does not incur any fuel costs or emission caps, it experiences resource variability and low capacity factors. The natural gas units can offer flexible dispatch and quick ramping capability in such cases for firming power system operations [25]. The uncertain characteristics of renewable energy introduce new challenges and additional costs to power system operations [26].

- Also a growing level of natural gas consumption by the electric power sector has increased the challenges associated with natural gas transmission network planning and operation. A pressure loss or interruption in natural gas pipelines could lead to the loss of multiple natural gas-fired generators and raise the power system security concerns. On the contrary, variable renewable energy units have zero fuel costs and relatively fixed operating cost when adequately distributed throughout a geographic region [27].

- Additional environmental regulations could further promote the use of clean and renewable energy in the near future [28]. Renewable energy will not be subject to certain environmental constraints which could introduce additional economics and security features to the operation of electric power systems. The proliferation of natural gas-fired and renewable energy units in electric power systems could be heightened even further as more stringent clean air requirements are enforced globally.

5.4 MANAGING THE RESOURCE UNCERTAINTY IN SCUC

The impact of uncertainty on the security of power system is of fundamental importance when multiple uncertain factors are integrated into power system operations. With the increased level of uncertainty in emerging power systems, deterministic SCUC models may offer unacceptable solutions in terms of cost and security risks. Incorporating certain types of uncertainty in SCUC could be very challenging in particular when considering the existing difficulties in calculating the deterministic SCUC solutions.

There are at least three solution techniques that have been proposed for managing uncertainties in SCUC, including stochastic programing (SP), robust optimization (RO), and chance-constrained optimization (CCO), with substantially different practical and computational requirements. This section briefly discusses the proposed models and the solution methodologies. Additional information is provided in References [29, 30].

5.4.1 SP-Based SCUC

SP is a well-known optimization technique for managing uncertainties in SCUC. In the SP approach, power system uncertainties are represented by a set of scenarios for the possible realization of various uncertainties. Usually, the scenario-based approach generates scenarios via presumed probability distribution functions for simulating uncertainties and each scenario is assigned a certain probability for its realization. The scenario generation methods include Monte Carlo (MC) sampling, moment matching principles, and methods motivated by stability analyses.

In SP, a large number of scenarios are needed for achieving an acceptable solution accuracy, which increases the size of the model, expands the computation burden, and limits the application to large-scale power systems. A survey of scenario tree algorithms is provided in Reference [31]. Thus, the scenario reduction technique [29] is usually adopted for reducing the scale of the stochastic model and the required computation effort, which aggregates close scenarios by measuring the probabilistic distance between scenarios and eliminates scenarios with very low probabilities. Low-discrepancy techniques, such as Latin Hypercube Sampling (LHS), are adopted for decreasing the variance of MC sampling. Other scenario reduction techniques may also be adopted including measuring the impact of each scenario on the objective by pre-solving individual single scenario problems, the target/moment matching which matches specified statistical properties, and the worst-case scenario probability study which assigns different sets of probabilities by experts and considers the worse scenarios.

The SP-based SCUC model is formulated as follows. The objective (5.32) is to minimize the costs of supplying the hourly load (which includes those of no-load, start-up, shutdown, and energy production) in the base case plus the expected corrective dispatch cost of scenarios, while satisfying various system and unit constraints. The sum of probabilities for all scenarios is equal to one, that is, $\sum_s p^s = 1$. By defining P^s_{ikt} as the power generation of unit i at time t of segment k in scenario s, which satisfies $P^s_{ikt} = P_{ikt} + \Delta P^s_{ikt}$, equation (5.32) can be equivalently converted to equation (5.33). The proposed SP-based SCUC model is a two-stage stochastic programming problem. The first-stage variables include UC decisions and generation dispatch of committed units in the base case in which the forecasts are made before scenarios are realized. The second-stage variables include corrective dispatch decisions for individual scenarios. It is noted that quick-start generating units may change their commitment status in scenarios in response to uncertainties. The formulation of quick-start units is included in the SCUC model [17,32]:

$$\text{Min} \sum_t \sum_i \left[N_i I_{it} + \text{SU}_{it} + \text{SD}_{it} + \sum_k c_{ik} P_{ikt} \right] + \sum_s p^s \sum_t \sum_i \left[\sum_k c_{ik} \Delta P^s_{ikt} \right] \tag{5.32}$$

$$\text{Min} \sum_t \sum_i \left[N_i I_{it} + \text{SU}_{it} + \text{SD}_{it} \right] + \sum_s p^s \sum_t \sum_i \left[\sum_k c_{ik} \Delta P^s_{ikt} \right] \tag{5.33}$$

subject to the following constraints:

Constraints for the base case   Base case constraints include the system load balance (5.2) and system reserve requirements (5.3). Generating unit constraints include ramping up/down limits as well as start-up/shutdown ramping for individual units in equation (5.4), minimum on and off time constraints in equation (5.5), real power generation limits in equation (5.6), constraints on spinning and operating reserves provided by individual generators in equation (5.7), constraint for wind farm (5.8), and constraints for batteries (5.9)–(5.14). Transmission network constraints include branch flow limits (5.15), which are enforced in the base case to guarantee the network security of power systems operation.

Constraints for each scenario Constraints for each scenario include the system load balance (5.34), generating unit capacity (5.35), and adjustment capabilities in the generating unit dispatch (5.36), which are restricted by the dispatch in the base case as well as the ramp up/down rates of unit i in scenario s ($R_i^{s,up}$ and $R_i^{s,dn}$). Ramping limits between successive intervals are not considered in each scenario because it is assumed such scenarios are independently operated. In addition, the proposed SP-based SCUC is an hourly model; thus it is reasonable to assume that there is enough time to adjust the base case operation status at hour t. Thus, ramping constraints would be required to guarantee the secure and economic transfer of the base case system state between two successive time intervals, and the hourly transfers from the base case to all scenarios, but not between successive intervals in a given scenario. Constraints (5.37) and (5.38)–(5.41) describe the limits on contingency dispatch for wind farms and batteries, respectively. Transmission network constraints (5.42) are enforced in each scenario for network security:

$$\sum_i P_{it}^s + \sum_w P_{wt}^s + \sum_v \left(P_{dc,vt}^s - P_{c,vt}^s\right) = D_t^s \tag{5.34}$$

$$P_i^{s,min} I_{it} \leq P_{it}^s \leq P_i^{s,max} I_{it}$$

$$P_{it}^s = \sum_k P_{ikt}^s \tag{5.35}$$

$$0 \leq P_{ikt}^s \leq P_{ik}^{s,max}$$

$$P_{it}^s - P_{it} \leq R_i^{s,up} I_{it}$$
$$P_{it} - P_{it}^s \leq R_i^{s,dn} I_{it} \tag{5.36}$$

$$0 \leq P_{wt}^s \leq P_{f,wt}^s \tag{5.37}$$

$$E_{vt}^s = E_{v(t-1)}^s - \left(P_{dc,vt}^s - \eta_v P_{c,vt}^s\right) \tag{5.38}$$

$$I_{c,vt} P_{c,v}^{s,min} \leq P_{c,vt}^s \leq I_{c,vt} P_{c,v}^{s,max} \tag{5.39}$$

$$I_{dc,vt} P_{dc,v}^{s,min} \leq P_{dc,vt}^s \leq I_{dc,vt} P_{dc,v}^{s,max} \tag{5.40}$$

$$E_v^{s,min} \leq E_{vt}^s \leq E_v^{s,max} \tag{5.41}$$

$$-PL_l^{s,max} \leq PL_{lt}^s = \sum_m LSF_l^{s,m} \left(\sum_{i:i\in U(m)} P_{it}^s + \sum_{w:w\in U(m)} P_{wt}^s \right.$$
$$\left. + \sum_{v:v\in U(m)} \left(P_{dc,vt}^s - P_{c,vt}^s\right) - D_{mt}^s\right) \leq PL_l^{s,max} \tag{5.42}$$

The above two-stage SP-based SCUC model is a large-scale, non-convex, NP-hard problem with a complicated solution. The Benders decomposition (BD) algorithm [33] is applied to large-scale MIP problems in power systems, including

SCUC and security constrained optimal power flow [7, 34–41]. BD would decompose the SP-based SCUC problem into a master problem and several tractable subproblems in each scenario. Other optimization methods, such as the scenario decomposition by relaxing non-anticipativity constraints and stochastic decomposition by relaxing demand constraints [29], are also adopted for solving the SCUC problem.

a. *Master problem in SCUC*

The master UC problem (5.43) minimizes the operation cost of the base case and the expected corrective dispatch cost of scenarios with respect to constraints (5.2)–(5.14):

$$\text{Min} \sum_t \sum_i [N_i I_{it} + SU_{it} + SD_{it}] + \sum_s p^s \sum_s \theta_t^s \qquad (5.43)$$

b. *Hourly network evaluation in the base case*

The subproblem (5.44) checks possible hourly network violations of the master UC solution in the base case. If the value of objective $\hat{s}_t$ in equation (5.44) is larger than the predefined threshold, a feasibility cut (5.45) will be utilized:

$$\text{Min } s_t$$

$$\text{S.t.} - s_t \le PL_l^{\max} - \sum_m LSF_l^m \left(\sum_{i:i\in U(m)} \hat{P}_{it} + \sum_{w:w\in U(m)} \hat{P}_{wt} \right.$$

$$\left. + \sum_{v:v\in U(m)} (\hat{P}_{dc,vt} - \hat{P}_{c,vt}) - D_{mt} \right) \qquad \lambda_{1,lt}$$

$$\qquad\qquad\qquad\qquad\qquad\qquad\qquad\qquad\qquad\qquad\qquad (5.44)$$

$$-s_t \le PL_l^{\max} + \sum_m LSF_l^m \left(\sum_{i:i\in U(m)} \hat{P}_{it} + \sum_{w:w\in U(m)} \hat{P}_{wt} \right.$$

$$\left. + \sum_{v:v\in U(m)} (\hat{P}_{dc,vt} - \hat{P}_{c,vt}) - D_{mt} \right) \qquad \lambda_{2,lt}$$

$$0 \le s_t$$

$$\hat{s}_t - \sum_l \left[(\lambda_{1,lt} - \lambda_{2,lt}) \sum_m LSF_l^m \left[\sum_{i:i\in U(m)} (P_{it} - \hat{P}_{it}) + \sum_{w:w\in U(m)} (P_{wt} - \hat{P}_{wt}) \right.\right.$$

$$\left.\left. + \sum_{v:v\in U(m)} [(P_{dc,vt} - P_{c,vt}) - (\hat{P}_{dc,vt} - \hat{P}_{c,vt})] \right] \right] \le 0 \qquad (5.45)$$

c. *Hourly feasibility check for each scenario*

The subproblem (5.46) checks possible hourly feasibility of the master UC solution in each scenario, in which λ and μ are dual variables of corresponding constraints in equation (5.46). If the value of objective function $\hat{S}_t^s$ in equation (5.46) is larger than the predefined threshold, a feasibility cut (5.47) will be utilized:

$$\text{Min}\quad S_t^s = s_t^s + s_{1t}^s + s_{2t}^s$$

$$\text{S.t.}\quad \sum_m \text{LSF}_l^{s,m}\left(\sum_{i:i\in U(m)} P_{it}^s + \sum_{w:w\in U(m)} P_{wt}^s\right.$$

$$+ \sum_{v:v\in U(m)} \left(P_{dc,vt}^s - P_{c,vt}^s\right) - D_{mt}^s\right) - s_t^s \le \text{PL}_l^{s,\max}$$

$$- \sum_m \text{LSF}_l^{s,m}\left(\sum_{i:i\in U(m)} P_{it}^s + \sum_{w:w\in U(m)} P_{wt}^s\right.$$

$$+ \sum_{v:v\in U(m)} \left(P_{dc,vt}^s - P_{c,vt}^s\right) - D_{mt}^s\right) - s_t^s \le \text{PL}_l^{s,\max}$$

$$\sum_i P_{it}^s + \sum_w P_{wt}^s + \sum_v \left(P_{dc,vt}^s - P_{c,vt}^s\right) + s_{1t}^s - s_{2t}^s = P_{Dt}^s$$

$$E_v^{s,\min} \le E_{vt}^s \le E_v^{s,\max}$$

$$0 \le P_{wt}^s \le P_{f,wt}^s$$

$$P_{it}^s \le R_i^{s,up}\hat{I}_{it} + \hat{P}_{it} \qquad\qquad \lambda_{1,it}^s$$

$$-P_{it}^s \le R_i^{s,dn}\hat{I}_{it} - \hat{P}_{it} \qquad\qquad \lambda_{2,it}^s$$

$$P_{it}^s \le P_i^{s,\max}\hat{I}_{it} \qquad\qquad \mu_{1,it}^s$$

$$-P_{it}^s \le -P_i^{s,\min}\hat{I}_{it} \qquad\qquad \mu_{2,it}^s$$

$$P_{c,vt}^s \le \hat{I}_{c,vt}P_{c,v}^{s,\max} \qquad\qquad \mu_{1c,vt}^s$$

$$-P_{c,vt}^s \le -\hat{I}_{c,vt}P_{c,v}^{s,\min} \qquad\qquad \mu_{2c,vt}^s$$

$$P_{dc,vt}^s \le \hat{I}_{dc,vt}P_{dc,v}^{s,\max} \qquad\qquad \mu_{1dc,vt}^s$$

$$-P_{dc,vt}^s \le -\hat{I}_{dc,vt}P_{dc,v}^{s,\min} \qquad\qquad \mu_{2dc,vt}^s \qquad\qquad (5.46)$$

$$E_{vt}^s + \left(P_{dc,vt}^s - \eta_v P_{c,vt}^s\right) = \hat{E}_{v(t-1)} \qquad\qquad \lambda_{vt}^s$$

$$P_{it}^s - \sum_k P_{ikt}^s = 0$$

$$0 \le P_{ikt}^s \le P_{ik}^{s,\max}$$

$$0 \le s_t^s, s_{1t}^s, s_{2t}^s$$

$$
\hat{S}_t^s + \sum_i \left[\left(\hat{\lambda}_{1,it}^s - \hat{\lambda}_{2,it}^s \right)(P_{it} - \hat{P}_{it}) + \left(\hat{\lambda}_{1,it}^s R_i^{s,up} + \hat{\lambda}_{2,it}^s R_i^{s,dn} \right. \right.
$$

$$
+ \hat{\mu}_{1,it}^s P_i^{s,\max} - \hat{\mu}_{2,it}^s P_i^{s,\min} \left) (I_{it} - \hat{I}_{it}) \right]
$$

$$
+ \sum_v \left[\left(\hat{\mu}_{1c,vt}^s P_{c,v}^{s,\max} - \hat{\mu}_{2c,vt}^s P_{c,v}^{s,\min} \right)(I_{c,vt} - \hat{I}_{c,vt}) + \left(\hat{\mu}_{1dc,vt}^s P_{dc,v}^{s,\max} \right. \right.
$$

$$
\left. - \hat{\mu}_{2dc,vt}^s P_{dc,v}^{s,\min} \right)(I_{dc,vt} - \hat{I}_{dc,vt}) + \lambda_{vt}^s (E_{v(t-1)} - \hat{E}_{v(t-1)}) \right] \le 0 \qquad (5.47)
$$

d. *Optimality check for each scenario*

The optimality check subproblem (5.48) checks the optimality of master UC solution in each scenario. If the objective value $\hat{W}_t^S$ in equation (5.48) is larger than the corrective dispatch cost $\hat{\theta}_t^S$ obtained from the master problem, the optimality cut (5.49) will be utilized:

$$
\text{Min} \quad W_t^s = \sum_i \left[\sum_k c_{ik} P_{ikt}^s \right]
$$

$$
\text{S.t.} - PL_l^{s,\max} \le \sum_m LSF_l^{s,m} \left(\sum_{i:i \in U(m)} P_{it}^s + \sum_{w:w \in U(m)} P_{wt}^s \right.
$$

$$
\left. + \sum_{v:v \in U(m)} \left(P_{dc,vt}^s - P_{c,vt}^s \right) - D_{mt}^s \right) \le PL_l^{s,\max}
$$

$$
0 \le P_{wt}^s \le P_{f,wt}^s
$$

$$
P_{it}^s \le R_i^{s,up} \hat{I}_{it} + \hat{P}_{it} \qquad\qquad \lambda_{1,it}^s
$$

$$
-P_{it}^s \le R_i^{s,dn} \hat{I}_{it} - \hat{P}_{it} \qquad\qquad \lambda_{2,it}^s
$$

$$
P_{it}^s \le P_i^{s,\max} \hat{I}_{it} \qquad\qquad \mu_{1,it}^s
$$

$$
-P_{it}^s \le -P_i^{s,\min} \hat{I}_{it} \qquad\qquad \mu_{2,it}^s \qquad\qquad (5.48)
$$

$$
P_{c,vt}^s \le \hat{I}_{c,vt} P_{c,v}^{s,\max} \qquad\qquad \mu_{1c,vt}^s
$$

$$
-P_{c,vt}^s \le -\hat{I}_{c,vt} P_{c,v}^{s,\min} \qquad\qquad \mu_{2c,vt}^s
$$

$$
P_{dc,vt}^s \le \hat{I}_{dc,vt} P_{dc,v}^{s,\max} \qquad\qquad \mu_{1dc,vt}^s
$$

$$
-P_{dc,vt}^s \le -\hat{I}_{dc,vt} P_{dc,v}^{s,\min} \qquad\qquad \mu_{2dc,vt}^s
$$

$$
E_{vt}^s + \left(P_{dc,vt}^s - \eta_v P_{c,vt}^s \right) = E_{v(t-1)}^s \qquad\qquad \lambda_{vt}^s
$$

$$
E_v^{s,\min} \le E_{vt}^s \le E_v^{s,\max}
$$

$$
P_{it}^s - \sum_k P_{ikt}^s = 0
$$

$$
0 \le P_{ikt}^s \le P_{ik}^{s,\max}
$$

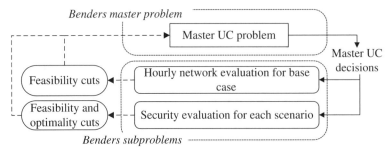

Figure 5.6   The SP-based SCUC solution approach.

$$
\begin{aligned}
\theta_t^s \geq \hat{W}_t^s &+ \sum_i \left[\left(\hat{\lambda}_{1,it}^s R_i^{s,up} + \hat{\lambda}_{2,it}^s R_i^{s,dn} + \hat{\mu}_{1,it}^s P_i^{s,max} - \hat{\mu}_{2,it}^s P_i^{s,min} \right) (I_{it} - \hat{I}_{it}) \right. \\
&+ \left(\hat{\lambda}_{1,it}^s - \hat{\lambda}_{2,it}^s \right) (P_{it} - \hat{P}_{it}) \right] \\
&+ \sum_v \left[\left(\hat{\mu}_{1,it}^s P_{c,v}^{s,max} - \hat{\mu}_{2c,it}^s P_{c,v}^{s,min} \right) (I_{c,vt} - \hat{I}_{c,vt}) \right. \\
&+ \left. \left(\hat{\mu}_{1dc,vt}^s P_{dc,v}^{s,max} - \hat{\mu}_{2dc,vt}^s P_{dc,v}^{s,min} \right) (I_{dc,vt} - \hat{I}_{dc,vt}) + \hat{\lambda}_{vt}^s (E_{v(t-1)} - \hat{E}_{v(t-1)}) \right]
\end{aligned}
$$

$$(5.49)$$

Figure 5.6 shows the flowchart of the SP-based SCUC solution. In this figure, the master UC problem (5.43) is solved first. Hourly UC and dispatch solutions are then passed on to the hourly network evaluation subproblems (5.44). The subproblems will examine the feasibility of the master solution. If a subproblem is infeasible, which violates the remaining constraints, a feasibility cut (5.45) will be generated and added to the next calculation of the master problem. The hourly UC and dispatch solutions are also passed on to the hourly security evaluation subproblem (5.46) and the optimality evaluation subproblem (5.48) in each scenario. If a scenario subproblem is infeasible, a corresponding feasibility cut (5.47) will be generated. If the optimal objective $\hat{W}_t^s$ is larger than the corrective dispatch cost $\hat{\theta}_t^s$, an optimality cut (5.49) will be generated and added to the next iteration of the master problem. The iterative process stops when the master solution satisfies all feasibility and optimality checks in subproblems.

5.4.2 RO-Based SCUC

RO is an alternative technique for managing uncertainties in the SCUC problem. The idea behind RO is to relegate the explicit knowledge on probability distributions and scenario samplings by using a deterministic uncertainty set. This uncertainty set considers a limited level of information on uncertain quantities, namely the mean value plus some estimate of the variance or a range of possible variations around the mean. Accordingly, the RO model seeks an optimal commitment and dispatch solution of the generating units for immunizing the solution against the worst economic condition (i.e., the highest minimum dispatch cost), which would preserve the power system against every possible realization of the presumed uncertainties in the chosen set.

The RO-based SCUC model is formulated as follows. The objective is to minimize the cost of supplying the hourly load (5.50) pertaining to the base case (5.2)–(5.15) and uncertainty (5.51)–(5.65) constraints. Here, we take load demand and wind uncertainties for discussing the RO-based SCUC model; other uncertainties can be considered similarly. Forecasted load demand D_{mt} and wind power generation $P_{f,wt}$ can introduce errors in the deterministic SCUC model. Thus, uncertainty intervals $[D_{mt} - \tilde{D}_{mt}, D_{mt} + \tilde{D}_{mt}]$ and $[P_{f,wt} - \tilde{P}_{wt}, P_{f,wt} + \tilde{P}_{wt}]$ are used to represent hourly variations. D_{mt}^{u} and $D_{f,wt}^{u}$ represent the set of load and wind realizations, which could take any values within possible intervals, that is, $D_{mt}^{u} \in [D_{mt} - \tilde{D}_{mt}, D_{mt} + \tilde{D}_{mt}]$ and $P_{f,wt}^{u} \in [P_{f,wt} - \tilde{P}_{wt}, P_{f,wt} + \tilde{P}_{wt}]$:

$$\operatorname*{Min}_{D_{mt}^{u} \in D, P_{f,wt}^{u} \in W, P_{it}, I_{it}} \sum_{t} \sum_{i} \left[\sum_{k} c_{ik} P_{ikt} + N_{i} I_{it} + \mathrm{SU}_{it} + \mathrm{SD}_{it} \right] \tag{5.50}$$

The power system balance constraint (5.51) would guarantee the system load balance when wind and load demand values are subject to uncertainties. Equations (5.52)–(5.54) describe generation dispatch limits of thermal units and wind farms under uncertainties. Thermal unit corrective dispatch adjustments in response to uncertainties are restricted by (5.53). Equations (5.55) and (5.56) represent hourly ramp up/down limits of thermal units. Equations (5.57)–(5.60) represent battery dispatch adjustment limits. Transmission network constraint (5.61) guarantees the network security under uncertainties. $P_{it}^{u}(D_{mt}^{u}, P_{f,wt}^{u})$ and $P_{wt}^{u}(D_{mt}^{u}, P_{f,wt}^{u})$ represent possible adjustments in thermal unit and wind farm dispatch which are in response to load demand and wind generation realizations D_{mt}^{u} and $P_{f,wt}^{u}$:

$$\sum_{i} P_{it}^{u}\left(D_{mt}^{u}, P_{f,wt}^{u}\right) + \sum_{v} \left[P_{dc,vt}^{u}\left(D_{mt}^{u}, P_{f,wt}^{u}\right) - P_{c,vt}^{u}\left(D_{mt}^{u}, P_{f,wt}^{u}\right)\right]$$
$$+ \sum_{w} P_{wt}^{u}\left(D_{mt}^{u}, P_{f,wt}^{u}\right) = \sum_{m} D_{mt}^{u} \tag{5.51}$$

$$P_{i}^{\min} I_{it} \le P_{it}^{u}\left(D_{mt}^{u}, P_{f,wt}^{u}\right) \le P_{i}^{\max} I_{it} \tag{5.52}$$

$$- R_{i}^{dn} I_{it} \le P_{it}^{u}\left(D_{mt}^{u}, P_{f,wt}^{u}\right) - P_{it} \le P_{i}^{up} I_{it} \tag{5.53}$$

$$0 \le P_{wt}^{u}\left(D_{mt}^{u}, P_{f,wt}^{u}\right) \le P_{f,wt}^{u} \tag{5.54}$$

$$P_{it}^{u}\left(D_{mt}^{u}, P_{f,wt}^{u}\right) - P_{i(t-1)}^{u}\left(D_{m(t-1)}^{u}, P_{f,w(t-1)}^{u}\right)$$
$$\le \mathrm{UR}_{i} I_{i(t-1)} + P_{i}^{\min}\left(I_{it} - I_{i(t-1)}\right) + P_{i}^{\max}(1 - I_{it}) \tag{5.55}$$

$$P_{i(t-1)}^{u}\left(D_{m(t-1)}^{u}, P_{f,w(t-1)}^{u}\right) - P_{it}^{u}\left(D_{mt}^{u}, P_{f,wt}^{u}\right)$$
$$\le \mathrm{DR}_{i} I_{it} + P_{i}^{\min}(I_{i(t-1)} - I_{it}) + P_{i}^{\max}(1 - I_{i(t-1)}) \tag{5.56}$$

$$E_{vt}^u\left(D_{mt}^u, P_{f,wt}^u\right) = E_{v(t-1)} - \left[P_{dc,vt}^u\left(D_{mt}^u, P_{f,wt}^u\right) - \eta_v P_{c,vt}^u\left(D_{mt}^u, P_{f,wt}^u\right)\right] \quad (5.57)$$

$$I_{c,vt}P_{c,v}^{\min} \le P_{c,vt}^u\left(D_{mt}^u, P_{f,wt}^u\right) \le I_{c,vt}P_{c,v}^{\max} \quad (5.58)$$

$$I_{dc,vt}P_{dc,v}^{\min} \le P_{dc,vt}^u\left(D_{mt}^u, P_{f,wt}^u\right) \le I_{dc,vt}P_{dc,v}^{\max} \quad (5.59)$$

$$E_v^{\min} \le E_{vt}^u\left(D_{mt}^u, P_{f,wt}^u\right) \le E_v^{\max} \quad (5.60)$$

$$-\mathrm{PL}_l^{\max} \le \sum_m \mathrm{LSF}_l^m \cdot \left(\sum_{i:i\in U(m)} P_{it}^u\left(D_{mt}^u, P_{f,wt}^u\right) + \sum_{v:v\in U(m)} \left[P_{dc,vt}^u\left(D_{mt}^u, P_{f,wt}^u\right)\right.\right.$$

$$\left.\left. - P_{c,vt}^u\left(D_{mt}^u, P_{f,wt}^u\right)\right] + \sum_{w:w\in U(m)} P_{wt}^u\left(D_{mt}^u, P_{f,wt}^u\right) - D_{mt}^u\right) \le \mathrm{PL}_l^{\max}$$

$$(5.61)$$

Two sets of constraints are introduced for controlling the conservatism of the RO model. The first one is the budget constraints (5.62) and (5.63), which are widely applied to RO models. These constraints indicate that total deviations of load and wind generation from their respective forecasts throughout the entire scheduling horizon cannot exceed the predefined budget limits Δ_d and Δ_w, which take values between 0 and NT. Higher Δ_d and Δ_w correspond to a larger uncertainty in D and W and a more conservative SCUC solution which is protected against a higher degree of uncertainty. When $\Delta_d = 0$ and $\Delta_w = 0$, the uncertainty sets are degraded to $D(D_{mt}, \tilde{D}_{mt}, 0) := \{D_{mt}\}$ and $W(P_{f,wt}, \tilde{P}_{wt}, 0) := \{P_{f,wt}\}$, which are equivalent to the deterministic case while neglecting all uncertainties. On the other hand, when $\Delta_d = \Delta_w = NT$, the uncertainty sets $D(D_{mt}, \tilde{D}_{mt}, NT)$ and $W(P_{f,wt}, \tilde{P}_{wt}, NT)$ are equal to the entire hyper cubes defined by the intervals for each D_{mt}^u and P_{wt}^u:

$$D(D_{mt}, \tilde{D}_{mt}, \Delta_d) := \left\{\sum_t \frac{|D_{mt}^u - D_{mt}|}{\tilde{D}_{mt}} \le \Delta_d, D_{mt}^u \in [D_{mt} - \tilde{D}_{mt}, D_{mt} + \tilde{D}_{mt}]\right\}$$

$$(5.62)$$

$$W(P_{f,wt}, \tilde{P}_{wt}, \Delta_w) := \left\{\sum_t \frac{|P_{f,wt}^u - P_{f,wt}|}{\tilde{P}_{wt}} \le \Delta_w, P_{f,wt}^u \in [P_{f,wt} - \tilde{P}_{wt}, P_{f,wt} + \tilde{P}_{wt}]\right\}$$

$$(5.63)$$

Historical load and wind data show that their variability may be correlated, that is, load and wind variability may follow certain patterns and the worst-case realizations of load and wind variability may not occur simultaneously. Thus, the proposed RO-based SCUC model considers load and wind variability correlations, which would eliminate unlike-to-happen scenarios and further limit the level of conservatism of the robust solution. Historical load and wind data are used to derive a moving average (MA) model of order one for simulating the linear correlation between load demand and wind generation, which is based on the theory of the least square regression. In addition, because load demands and wind generations are uncertain, the correlation between load and wind generation is formulated in terms that load demand is located in the interval with the mean of $a_{wdt} + b_{wdt}P^u_{f,wt}$ and a certain confidence level (i.e., 95%), as shown in equation (5.64). As the number of historical load and wind samples is limited, $[D^u_{mt} - (a_{wdt} + b_{wdt}P^u_{f,wt})]/\sigma_{wdt}$ would follow the t-distribution. If load and wind are strongly linearly correlated, that is, most P^u_{dt} are close to $(a_{wdt} + b_{wdt}P^u_{f,wt})$, the variance σ^2_{wdt} will be small according to equation (5.65) and in turn the interval $[(a_{wdt} + b_{wdt}P^u_{f,wt}) - t^{0.975}_{NS-1}\sigma_{wdt}, (a_{wdt} + b_{wdt}P^u_{f,wt}) + t^{0.975}_{NS-1}\sigma_{wdt}]$ (i.e., 95% confidence level) will be narrow. Otherwise, equation (5.64) will provide a wider interval for representing load and wind generation uncertainty correlations. a_{wdt} and b_{wdt} are parameters to represent the correlation of wind farm w and load d at time t. $\overline{P}_{wt}$ is the mean value of historical wind generation samples of wind farm w at time t. $\overline{P}_{dt}$ is the mean value of historical load samples for load d at time t. σ_{wdt} is the variance of the deviation of historical wind and load data from the approximated linear correlation representation:

$$\left(a_{wdt} + b_{wdt}P^u_{f,wt}\right) - t^{0.975}_{NS-1}\sigma_{wdt} \leq D^u_{mt} \leq \left(a_{wdt} + b_{wdt}P^u_{f,wt}\right) + t^{0.975}_{NS-1}\sigma_{wdt} \quad (5.64)$$

$$a_{wdt} = \overline{D}_{mt,s} - b_{wdt}\overline{P}_{wt,s}$$

$$b_{wdt} = \frac{NS\sum_s(P_{wt,s}D_{mt,s}) - \sum_s P_{wt,s}\sum_s D_{mt,s}}{NS\sum_s(P_{wt,s}P_{wt,s}) - \sum_s P_{wt,s}\sum_s P_{wt,s}} \quad (5.65)$$

$$\sigma^2_{wdt} = \frac{\sum_s(D_{mt,s} - \overline{D}_{mt,s})^2 - b_{wdt}\sum_s(D_{mt,s} - \overline{D}_{mt,s})(P_{wt,s} - \overline{P}_{wt,s})}{NS - 2}$$

Budget constraints (5.62) and (5.63) describe the temporal relationship of load and wind uncertainties in terms of uncertainty budget limits throughout the scheduling horizon, which need to be carefully tuned in order to control the conservatism of the robust solution. Different from the budget constraints, load and wind uncertainty correlation constraints (5.64) and (5.65) describe the relationship between load and wind generation uncertainties in each hour for enhancing the conservatism control of the robust solution. In addition, unlike budget constraints (5.62) and (5.63) which need heuristic parameter tuning, load and wind uncertainty correlations (5.64) and (5.65) can be accurately derived from historical data.

The proposed RO-based SCUC model (5.50), (5.2)–(5.15), and (5.51)–(5.65) is a large-scale, non-convex, NP-hard problem with interval parameters. Benders decomposition is used to decompose the original problem into a master UC problem for the base case and tractable subproblems for the base case network evaluation and the security checking for uncertainty intervals. The detailed formulations for the master problem and subproblems, as well as the detailed algorithm steps are described as follows.

(a) *Master UC problem*

The master UC problem minimizes the operation cost of the base case (5.66) subject to constraints (5.2)–(5.14) and all Benders cuts obtained so far:

$$\text{Min} \sum_i \sum_t \left[\sum_k c_{ik} P_{ikt} + N_i I_{it} + \text{SU}_{it} + \text{SD}_{it} \right] \tag{5.66}$$

(b) *Hourly network evaluation for the base case*

The hourly network evaluation subproblem (5.67) checks possible network violations of the master UC solution in the base case, where $\lambda_{1,l,t}$ and $\lambda_{2,l,t}$ are dual variables corresponding to the two transmission capacity limitation constraints. If the objective value of (5.67) is larger than the predefined threshold, a feasibility cut (5.68) is generated:

$$\text{Min } s_t$$

$$\text{S.t.} - s_t \le \text{PL}_l^{\max} - \sum_m \text{LSF}_l^m \left(\sum_{i:i\in U(m)} \hat{P}_{it} + \sum_{v:v\in U(m)} [\hat{P}_{dc,vt} - \hat{P}_{c,vt}] \right.$$

$$\left. + \sum_{w:w\in U(m)} \hat{P}_{wt} - D_{mt} \right) \lambda_{1,lt}$$

$$\tag{5.67}$$

$$-s_t \le \text{PL}_l^{\max} + \sum_m \text{LSF}_l^m \left(\sum_{i:i\in U(m)} \hat{P}_{it} + \sum_{v:v\in U(m)} [\hat{P}_{dc,vt} - \hat{P}_{c,vt}] \right.$$

$$\left. + \sum_{w:w\in U(m)} \hat{P}_{wt} - D_{mt} \right) \lambda_{2,lt}$$

$$0 \le s_t$$

$$\hat{s}_t - \sum_l \left[\left(\lambda_{1,lt} - \lambda_{2,lt} \right) \sum_m \text{LSF}_l^m \left[\sum_{i:i\in U(m)} (P_{it} - \hat{P}_{it}) \right. \right.$$

$$+ \sum_{v:v\in U(m)} [(P_{dc,vt} - P_{c,vt}) - (\hat{P}_{dc,vt} - \hat{P}_{c,vt})]$$

$$\left. \left. + \sum_{w:w\in U(m)} (P_{wt} - \hat{P}_{wt}) \right] \right] \le 0 \tag{5.68}$$

(c) *Security evaluation for uncertainty intervals*

This is the key part of the RO-based SCUC model, which identifies possible violations when wind and load vary within their given intervals of uncertainty (5.69). If any violation exists, feasibility Benders cuts will be generated and fed back to the master UC problem for seeking new UC solutions that would alleviate transmission security violations for the entire uncertainty intervals. Equation (5.69) can be solved sequentially via two substeps. The first substep identifies uncertain wind and load quantity realizations which would lead to the largest minimum violation. If the largest minimum violation is smaller than the predefined threshold, the security evaluation for uncertainty intervals will be accepted. Otherwise, the corresponding realizations D_{mt}^{worst} and $P_{f,wt}^{\text{worst}}$ will be passed on to the second substep to generate Benders cuts and feedback them to the master problem:

$$\min_{D_{mt}^u \in D, P_{f,wt}^u \in W} \sum_t \left(\sum_l s_{1,lt} + s_{2t} + s_{3t} \right)$$

$$\text{S.t.} \sum_m \text{LSF}_l^m \left(\sum_{i:i \in U(m)} P_{it}^u \left(D_{mt}^u, P_{f,wt}^u \right) + \sum_{v:v \in U(m)} \left[P_{dc,vt}^u \left(D_{mt}^u, P_{f,wt}^u \right) \right. \right.$$

$$\left. \left. - P_{c,vt}^u \left(D_{mt}^u, P_{f,wt}^u \right) \right] + \sum_{w:w \in U(m)} P_{wt}^u \left(D_{mt}^u, P_{f,wt}^u \right) - D_{mt}^u \right) - s_{1,lt} \le \text{PL}_l^{\max}$$

$$- \sum_m \text{LSF}_l^m \cdot \left(\sum_{i:i \in U(m)} \Gamma_{it}^u \left(D_{mt}^u, \Gamma_{f,wt}^u \right) + \sum_{v:v \in U(m)} \left[\Gamma_{dc,vt}^u \left(D_{mt}^u, \Gamma_{f,wt}^u \right) \right. \right.$$

$$\left. \left. - P_{c,vt}^u \left(D_{mt}^u, P_{f,wt}^u \right) \right] + \sum_{w:w \in U(m)} P_{wt}^u \left(D_{mt}^u, P_{f,wt}^u \right) - D_{mt}^u \right) - s_{1,lt} \le \text{PL}_l^{\max}$$

$$\sum_i P_{it}^u \left(D_{mt}^u, P_{f,wt}^u \right) + \sum_v \left[P_{dc,vt}^u \left(D_{mt}^u, P_{f,wt}^u \right) - P_{c,vt}^u \left(D_{mt}^u, P_{f,wt}^u \right) \right]$$

$$+ \sum_w P_{wt}^u \left(D_{mt}^u, P_{f,wt}^u \right) + s_{2t} - s_{3t} = \sum_m D_{mt}^u \qquad (5.69)$$

Constraints (5.53)–(5.61) and (5.63)–(5.66) with given base case solutions

$$0 \le s_{1,lt}, s_{2t}, s_{3t}$$

(c.1) *Identify the largest minimum violation*

The idea of identifying the largest minimum violation when wind and load quantities vary within their uncertainty intervals is to transform equation (5.69) into an equivalent max–min optimization problem (5.70) for calculating the largest minimum violation. $\lambda_{1,lt}, \lambda_{2,lt}, \lambda_{3t}, \lambda_{4,wt}, \lambda_{5,it}, \lambda_{6,it}, \lambda_{7,it},$

$\lambda_{8,it}$, $\lambda_{9,it}$, $\lambda_{10,it}$, $\lambda_{11,vt}$, $\lambda_{12,vt}$, $\lambda_{13,vt}$, $\lambda_{14,vt}$, $\lambda_{15,vt}$, $\lambda_{16,vt}$, and $\lambda_{17,vt}$ are corresponding dual variables of constraints in equation (5.70):

$$\underset{D^u_{mt},P^u_{f,wt}}{\text{Max}} \quad \underset{s_{1,lt},s_{2t},s_{3t},P^u_{it},P^u_{vt},P^u_{wt}}{\text{Min}} \sum_t \left(\sum_l s_{1,lt} + s_{2t} + s_{3t} \right)$$

$$\text{S.t.} \sum_m \text{LSF}^m_l \left(\sum_{i:i\in U(m)} P^u_{it} + \sum_v \left[P^u_{dc,vt} - P^u_{c,vt} \right] \right.$$

$$\left. + \sum_{w:w\in U(m)} P^u_{wt} - D^u_{mt} \right) - s_{1,lt} \leq \text{PL}^{\max}_l \qquad \lambda_{1,lt}$$

$$- \sum_m \text{LSF}^m_l \left(\sum_{i:i\in U(m)} P^u_{it} + \sum_v \left[P^u_{dc,vt} - P^u_{c,vt} \right] \right.$$

$$\left. + \sum_{w:w\in U(m)} P^u_{wt} - D^u_{mt} \right) - s_{1,lt} \leq \text{PL}^{\max}_l \qquad \lambda_{2,lt}$$

$$\sum_i P^u_{it} + \sum_v \left[P^u_{dc,vt} - P^u_{c,vt} \right] + \sum_w P^u_{wt} + s_{2t} - s_{3t} = \sum_m D^u_{mt} \qquad \lambda_{3t}$$

$$0 \leq P^u_{wt} \leq P^u_{f,wt} \qquad \lambda_{4,wt}$$

$$P^{\min}_i \hat{I}^b_{it} \leq P^u_{it} \leq P^{\max}_i \hat{I}^b_{it} \qquad \lambda_{6,it}, \lambda_{5,it}$$

$$-R^{dn}_i \hat{I}^b_{it} \leq P^u_{it} - \hat{P}^b_{it} \leq R^{up}_i \hat{I}^b_{it} \qquad \lambda_{8,it}, \lambda_{7,it}$$

$$P^u_{it} - P^u_{i(t-1)} \leq \text{UR}_i \hat{I}^b_{i(t-1)} + P^{\min}_i \left(\hat{I}^b_{it} - \hat{I}^b_{i(t-1)} \right) + P^{\max}_i \left(1 - \hat{I}^b_{it} \right) \qquad \lambda_{9,it}$$

$$P^u_{i(t-1)} - P^u_{it} \leq \text{DR}_i \hat{I}^b_{it} + P^{\min}_i \left(\hat{I}^b_{i(t-1)} - \hat{I}^b_{it} \right) + P^{\max}_i \left(1 - \hat{I}^b_{i(t-1)} \right) \qquad \lambda_{10,it}$$

$$E^u_{vt} + \left[P^u_{dc,vt} - \eta_v P^u_{c,vt} \right] = \hat{E}_{v(t-1)} \qquad \lambda_{11,vt}$$

$$\hat{I}_{c,vt} P^{\min}_{c,v} \leq P^u_{c,vt} \leq \hat{I}_{c,vt} P^{\max}_{c,v} \qquad \lambda_{13,vt}, \lambda_{12,vt}$$

$$\hat{I}_{dc,vt} P^{\min}_{dc,v} \leq P^u_{dc,vt} \leq \hat{I}_{dc,vt} P^{\max}_{dc,v} \qquad \lambda_{15,vt}, \lambda_{14,vt}$$

$$E^{\min}_v \leq E^u_{vt} \leq E^{\max}_v \qquad \lambda_{17,vt}, \lambda_{16,vt}$$

Constraints (5.63)–(5.66)

$$0 \leq s_{1,lt}, s_{2t}, s_{3t} \qquad (5.70)$$

By applying the duality theory of the LP problem, the inner minimization problem of equation (5.70) will be transferred into a maximization problem. Thus, equation (5.70) can be converted into a single-level bilinear optimization problem (5.71). Several methods are applied to solve the bilinear optimization problem (5.71), including interior-point methods [42] and effective heuristic methods such as the outer approximation (OA) method [43]. The OA method is used here to illustrate how the bilinear optimization problem (5.71) is solved. In OA, the bilinear term in the objective is linearly approximated around the current solution point and added to the OA formulation. In order to verify the quality of the final solution, the bilinear optimization problem (5.71) is solved with different initial values for D_{mt}^u and $P_{f,wt}^u$, that is, uncertainty intervals $[D_{mt} - \tilde{D}_{mt}, D_{mt} + \tilde{D}_{mt}]$ and $[P_{f,wt}^b - \tilde{P}_{wt}, P_{f,wt}^b + \tilde{P}_{wt}]$ are equally divided into n segments to derive $(n+1)$ initial D_{mt}^u and $P_{f,wt}^u$ values, and the bilinear optimization problem (5.71) is solved $(n+1)^2$ times in parallel with various combinations of initial values for exploring the optimal solution. $\text{LSF}_l^{m(i)}$ and $\text{LSF}_l^{m(w)}$ represent the linear sensitivity of flow on line l to power injection at buses where the thermal unit i and the wind farm w are located:

$$
\begin{aligned}
\max_{P_{dt}^u, P_{f,wt}^u} \sum_{t=1}^{NT} &\left\{ \begin{aligned}
&\sum_l (\lambda_{1,lt} - \lambda_{2,lt}) \left(\sum_m \text{LSF}_l^m D_{mt}^u \right) + \sum_l (\lambda_{1,lt} + \lambda_{2,lt}) \text{PL}_l^{\max} \\
&+ \lambda_{3,t} \sum_m D_{mt}^u + \sum_w \lambda_{4,wt} P_{f,wt}^u + \\
&\sum_i \left[\lambda_{5,it} \left(P_i^{\max} \hat{I}_{it}^b \right) - \lambda_{6,it} \left(P_i^{\min} \hat{I}_{it}^b \right) \right] \\
&+ \sum_i \lambda_{7,it} \left(\hat{P}_{it}^b + R_l^{up} \hat{I}_{it}^b \right) - \sum_i \lambda_{8,it} \left(\hat{P}_{it}^b - R_l^{dn} \hat{I}_{it}^b \right) + \\
&\sum_v \left[\lambda_{11,vt} \hat{E}_{v(t-1)} + \lambda_{12,vt} \left(\hat{I}_{c,vt} P_{c,v}^{\max} \right) - \lambda_{13,vt} \left(\hat{I}_{c,vt} P_{c,v}^{\min} \right) \right. \\
&+ \lambda_{14,vt} \left(\hat{I}_{dc,vt} P_{dc,vt}^{\max} \right) - \lambda_{15,vt} \left(\hat{I}_{dc,vt} P_{dc,vt}^{\min} \right) \\
&\left. + \lambda_{16,vt} E_v^{\max} - \lambda_{17,vt} E_v^{\min} \right) \right]
\end{aligned} \right\} \\
+ \sum_{t=2}^{NT} &\left\{ \sum_i \lambda_{9,it} \left[\text{UR}_i \hat{I}_{i(t-1)}^b + P_i^{\min} \left(\hat{I}_{it}^b - \hat{I}_{i(t-1)}^b \right) + P_i^{\max} \left(1 - \hat{I}_{it}^b \right) \right] \right. \\
&\left. + \sum_i \lambda_{10,it} \left[\text{DR}_i \hat{I}_{it}^b + P_i^{\min} \left(\hat{I}_{i(t-1)}^b - \hat{I}_{it}^b \right) + P_i^{\max} \left(1 - \hat{I}_{i(t-1)}^b \right) \right] \right\}
\end{aligned}
$$

$$\text{S.t.} \quad -\lambda_{1,lt} - \lambda_{2,lt} \leq 1$$

$$-1 \leq \lambda_{3,t} \leq 1$$

$$\sum_l \left(\lambda_{1,lt} - \lambda_{2,lt} \right) \text{LSF}_l^{m(i)} + \lambda_{3,t} + \lambda_{5,it} - \lambda_{6,it} + \lambda_{7,it} - \lambda_{8,it}$$

$$-\lambda_{9,i(t+1)} + \lambda_{10,i(t+1)} \leq 0, \quad t = 1$$

$$\sum_l \left(\lambda_{1,lt} - \lambda_{2,lt}\right) \mathrm{LSF}_l^{m(i)} + \lambda_{3,t} + \lambda_{5,it} - \lambda_{6,it} + \lambda_{7,it} - \lambda_{8,it}$$

$$+\lambda_{9,it} - \lambda_{9,i(t+1)} - \lambda_{10,it} + \lambda_{10,i(t+1)} \le 0, \quad \forall t = 2, \dots, NT - 1$$

$$\sum_l \left(\lambda_{1,lt} - \lambda_{2,lt}\right) \mathrm{LSF}_l^{m(i)} + \lambda_{3,t} + \lambda_{5,it} - \lambda_{6,it} + \lambda_{7,it} - \lambda_{8,it}$$

$$+\lambda_{9,it} - \lambda_{10,it} \quad t = NT$$

$$\sum_l \left(\lambda_{1,lt} - \lambda_{2,lt}\right) \mathrm{LSF}_l^{m(w)} + \lambda_{3,t} + \lambda_{4,wt} \le 0$$

$$-\sum_l \left(\lambda_{1,lt} - \lambda_{2,lt}\right) \mathrm{LSF}_l^{m(v)} - \lambda_{3,t} - \eta_v \cdot \lambda_{11,vt} + \lambda_{12,vt} - \lambda_{13,vt} \le 0$$

$$\sum_l \left(\lambda_{1,lt} - \lambda_{2,lt}\right) \mathrm{LSF}_l^{m(v)} + \lambda_{3,t} + \lambda_{11,vt} + \lambda_{14,vt} - \lambda_{15,vt} \le 0$$

$$\lambda_{11,vt} + \lambda_{16,vt} - \lambda_{17,vt} \le 0$$

Constraints   (5.63) − (5.66)

$$\lambda_{1,lt}, \lambda_{2,lt}, \lambda_{4,lt}, \lambda_{5,lt}, \lambda_{6,lt}, \lambda_{7,lt}, \lambda_{8,lt}, \lambda_{9,lt}, \lambda_{10,lt}, \lambda_{12,vt}, \lambda_{13,vt}, \lambda_{14,vt},$$

$$\lambda_{15,vt}, \lambda_{16,vt}, \lambda_{17,vt} \le 0, \quad \lambda_{3,t}, \lambda_{11,vt} \quad \text{unlimited} \tag{5.71}$$

(c.2) *Generate Benders cuts corresponding to wind and load realizations with the largest minimum violation*

If the largest minimum security violation obtained in equation (5.71) exceeds the predefined threshold, the hourly security checking subproblem (5.72) will generate the feasibility Benders cut (5.73) with respect to the largest minimum violation realizations D_{mt}^{worst} and $P_{f,wt}^{\mathrm{worst}}$ obtained in (5.71). The cut (5.73) is fed back to the master UC problem for seeking robust UC solutions that would alleviate transmission security violations for the entire uncertainty intervals:

$$\mathrm{Min} \quad s = \sum_t \left(\sum_l s_{1,l,t} + s_{2t} + s_{3t} \right)$$

$$\text{S.t.} \quad \sum_m \mathrm{LSF}_l^m \begin{pmatrix} \displaystyle\sum_{i:i \in U(m)} P_{it}^u \left(D_{mt}^{\mathrm{worst}}, P_{f,wt}^{\mathrm{worst}}\right) \\[2mm] + \displaystyle\sum_{v:v \in U(m)} \left[P_{dc,vt}^u \left(D_{mt}^{\mathrm{worst}}, P_{f,wt}^{\mathrm{worst}}\right) \right. \\[2mm] \left. - P_{c,vt}^u \left(D_{mt}^{\mathrm{worst}}, P_{f,wt}^{\mathrm{worst}}\right) \right] \\[2mm] + \displaystyle\sum_{w:w \in U(m)} P_{wt}^u \left(D_{mt}^{\mathrm{worst}}, P_{f,wt}^{\mathrm{worst}}\right) - D_{mt}^{\mathrm{worst}} \end{pmatrix} - s_{1,lt} \le \mathrm{PL}_l^{\mathrm{max}}$$

$$- \sum_m \text{LSF}_l^m \begin{pmatrix} \sum_{i:i \in U(m)} P_{it}^u \left(D_{mt}^{\text{worst}}, P_{f,wt}^{\text{worst}} \right) \\ + \sum_{v:v \in U(m)} \left[P_{dc,vt}^u \left(D_{mt}^{\text{worst}}, P_{f,wt}^{\text{worst}} \right) \\ - P_{c,vt}^u \left(D_{mt}^{\text{worst}}, P_{f,wt}^{\text{worst}} \right) \right] \\ + \sum_{w:w \in U(m)} P_{wt}^u \left(D_{mt}^{\text{worst}}, P_{f,wt}^{\text{worst}} \right) - D_{mt}^{\text{worst}} \end{pmatrix} - s_{1,lt} \le \text{PL}_l^{\max}$$

$$\sum_i P_{it}^u \left(D_{mt}^{\text{worst}}, P_{f,wt}^{\text{worst}} \right) + \sum_w P_{wt}^u \left(D_{mt}^{\text{worst}}, P_{f,wt}^{\text{worst}} \right)$$

$$+ \sum_v \left[P_{dc,vt}^u \left(D_{mt}^{\text{worst}}, P_{f,wt}^{\text{worst}} \right) - P_{c,vt}^u \left(D_{mt}^{\text{worst}}, P_{f,wt}^{\text{worst}} \right) \right] + s_{2t} - s_{3t} = \sum_m D_{mt}^{\text{worst}}$$

$$0 \le P_{wt}^u \left(D_{mt}^{\text{worst}}, P_{f,wt}^{\text{worst}} \right) \le P_{f,wt}^{\text{worst}}$$

$$\hat{P}_{it}^b - R_i^{dn} \hat{I}_{it}^b \le P_{it}^u \left(D_{mt}^{\text{worst}}, P_{f,wt}^{\text{worst}} \right) \le \hat{P}_{it}^b + R_i^{dn} \hat{I}_{it}^b \quad \lambda_{2,it}, \lambda_{1,it}$$

$$P_i^{\min} \hat{I}_{it}^b \le P_{it}^u \left(D_{mt}^{\text{worst}}, P_{f,wt}^{\text{worst}} \right) \le P_i^{\max} \hat{I}_{it}^b \quad \mu_{2,it}, \mu_{1,it}$$

$$P_{it}^u \left(D_{mt}^{\text{worst}}, P_{f,wt}^{\text{worst}} \right) - P_{i(t-1)}^u \left(D_{m(t-1)}^{\text{worst}}, P_{f,w(t-1)}^{\text{worst}} \right)$$

$$\le \text{UR}_i \hat{I}_{i(t-1)}^b + P_i^{\min} \left(\hat{I}_{it}^b - \hat{I}_{i(t-1)}^b \right) + P_i^{\max} \left(1 - \hat{I}_{it}^b \right) \quad \eta_{1,it}$$

$$P_{i(t-1)}^u \left(D_{m(t-1)}^{\text{worst}}, P_{f,w(t-1)}^{\text{worst}} \right) - P_{it}^u \left(D_{mt}^{\text{worst}}, P_{f,wt}^{\text{worst}} \right)$$

$$< \text{DR}_i \hat{I}_{ii}^b + P_i^{\min} \left(\hat{I}_{i(t-1)}^b - \hat{I}_{ii}^b \right) + P_i^{\max} \left(1 - \hat{I}_{l(t-1)}^b \right) \quad \eta_{2,it}$$

$$\hat{I}_{c,vt} P_{c,v}^{\min} \le P_{c,vt}^u \left(D_{mt}^{\text{worst}}, P_{f,wt}^{\text{worst}} \right) \le \hat{I}_{c,vt} P_{c,v}^{\max} \quad \lambda_{2,vt}, \lambda_{1,vt}$$

$$\hat{I}_{dc,vt} P_{dc,v}^{\min} \le P_{dc,vt}^u \left(D_{mt}^{\text{worst}}, P_{f,wt}^{\text{worst}} \right) \le \hat{I}_{dc,vt} P_{dc,v}^{\max} \quad \lambda_{4,vt}, \lambda_{3,vt}$$

$$E_{vt}^u \left(D_{mt}^{\text{worst}}, P_{f,wt}^{\text{worst}} \right) + \left[P_{dc,vt}^u \left(D_{mt}^{\text{worst}}, P_{f,wt}^{\text{worst}} \right) \right.$$

$$\left. - \eta_v P_{c,vt}^u \left(D_{mt}^{\text{worst}}, P_{f,wt}^{\text{worst}} \right) \right] = \hat{E}_{v,(t-1)} \quad \lambda_{5,vt}$$

$$E_v^{\min} \le E_{vt}^u \left(D_{mt}^{\text{worst}}, P_{f,wt}^{\text{worst}} \right) \le E_v^{\max}$$

$$0 \le s_{1,lt}, s_{2t}, s_{3t} \tag{5.72}$$

$$\hat{s} + \sum_{t=1}^{NT} \sum_i \left[\left(\hat{\lambda}_{1,it} R_i^{up} + \hat{\lambda}_{2,it} R_i^{dn} + \hat{\mu}_{1,it} P_i^{\max} - \hat{\mu}_{2,it} P_i^{\min} \right) \left(I_{it}^b - \hat{I}_{it}^b \right) \right.$$

$$\left. + \left(\hat{\lambda}_{1,it} - \hat{\lambda}_{2,it} \right) \left(P_{it}^b - \hat{P}_{it}^b \right) \right]$$

$$+ \sum_{t=1}^{NT} \sum_v \left[\left(\hat{\lambda}_{1,it} P_{c,v}^{\max} - \hat{\lambda}_{2,it} P_{c,v}^{\min} \right) \left(I_{c,vt} - \hat{I}_{c,vt} \right) \right.$$

$$\left. + \left(\hat{\lambda}_{3,vt} P_{dc,v}^{\max} - \hat{\lambda}_{4,vt} P_{dc,v}^{\min} \right) \left(I_{dc,vt} - \hat{I}_{dc,vt} \right) + \hat{\lambda}_{5,vt} \left(E_{vt} - \hat{E}_{vt} \right) \right]$$

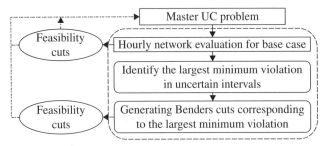

Figure 5.7 Flowchart of the RO-based SCUC solution methodology.

$$+ \sum_{t=2}^{NT} \sum_{i} \left[\hat{\eta}_{1,it}\left(\text{UR}_i - P_i^{\min}\right) + \hat{\eta}_{2,it}\left(P_i^{\min} - P_i^{\max}\right)\right]\left(I_{i(t-1)}^b - \hat{I}_{i(t-1)}^b\right)$$

$$+ \sum_{t=2}^{NT} \sum_{i} \left[\hat{\eta}_{1,it}\left(P_i^{\min} - P_i^{\max}\right) + \hat{\eta}_{2,it}\left(\text{DR}_i - P_i^{\min}\right)\right]\left(I_{it}^b - \hat{I}_{it}^b\right) \leq 0 \quad (5.73)$$

As shown in Figure 5.7, the RO-based SCUC model is solved by the following steps:

Step 1: The master UC problem for the base case is solved, which minimizes the operation cost for the base case (5.66) subject to constraints (5.2)–(5.14). The optimal solutions $\hat{P}_{it}$ and $\hat{I}_{it}$ are passed on to the base case hourly network evaluation subproblem in Step 2 and the security evaluation subproblem for uncertainty intervals in Step 3.

Step 2: The base case hourly transmission network evaluation subproblem (5.67) checks the possible transmission network violations based on the master UC solution for the base case. If the objective value (5.67) is larger than the predefined threshold, a feasibility cut (5.68) will be generated.

Step 3: This step, which includes two substeps, checks possible violations when wind and load vary within their uncertainty intervals. The first substep solves equation (5.71) for identifying the worst wind and load realizations D_{mt}^{worst} and $P_{f,wt}^{\text{worst}}$ that would lead to the largest minimum transmission security violation, when the hourly wind energy and load vary with their intervals. If the largest minimum violation is larger than the predefined threshold, the second substep solves the hourly transmission security check subproblem (5.72) corresponding to the worst wind and load realizations D_{mt}^{worst} and $P_{f,wt}^{\text{worst}}$ and generates the feasibility Benders cut (5.73).

Step 4: Feasibility cuts (5.68) and (5.73) generated in Steps 2 and 3 are fed back to the master problem in Step 1 for seeking robust UC solutions that would alleviate transmission security violations. The iterative procedure will stop when the master solution satisfies all transmission security violation checks, that is, no more feasibility cuts are generated in Steps 2 and 3.

5.4.3 Chance-Constrained Optimization-Based SCUC

Chance-constrained optimization is another viable approach for handling uncertainties in the hourly SCUC problems, in which temporal constraints can be violated with a predefined level of probability. Thus, CCO offers a good alternative for decision makers to understand and manage the trade-off between cost and robustness, by leveraging the probability that certain constraints may be violated. CCO matches the nature of the SCUC operation in essence that one may not actually be able to guarantee that transmission security constraints (such as system load balance or transmission capacity) will never be violated. Rather, one has to provide SCUC solutions which are "reasonably feasible" (i.e., offering limited load shedding) under all except the most unlikely scenarios.

In CCO, the safety level can be set for individual constraints, representing an individual CCO (ICCO), or for the entire system resulting in a joint CCO (JCCO). While the ICCO is usually less robust than the JCCO, the latter is significantly more difficult to solve. Usually, such chance constraints are non-convex and often intractable, which make the approach computationally taxing. The solution to the CCO-based SCUC problem can be obtained by sampling scenarios to approximate the true distribution of random variables or converting to a sequence of deterministic equivalents using techniques such as the central limit theorem, which converge the approximated solution to that of the original CCO-based SCUC model [44]. The SCUC problem scale and the computation burden of the two options are comparable to those of scenario-based and interval optimization approaches, respectively. Considering hourly load forecast errors, the UC problem is formulated as a CCO problem in which the hourly load is fully supplied with a high probability level. The chance constraints are then replaced by a set of probability constraints, which are converted to a set of deterministic linear equivalents. In Reference [45], the hourly UC problem is formulated as a chance-constrained two-stage stochastic programming problem and solved by a combined sample average approximation algorithm, which guarantees that a large portion of the uncertain hourly wind power generation will be utilized with a high probability.

The CCO-based SCUC model is formulated as follows. The objective is to minimize the cost of supplying the hourly load in the base case (5.1), while satisfying various system and generating unit constraints in the base case (5.2), (5.4)–(5.14), and (5.74)–(5.76) and contingencies (5.17)–(5.23) and (5.77)–(5.79) when various system uncertainties are considered. The system hourly reserve requirement (5.3) and the DC network evaluation constraint (5.15) for the base case are replaced by chance constraints (5.74)–(5.75) and (5.77) with preassigned probability levels, for satisfying the economics and reliability criteria of power systems with uncertainties. The day-ahead scheduling provides sufficient reserves to accommodate various system uncertainties, which is modeled as a chance constraint (5.74). Constraint (5.74) indicates that the total generation and reserve must satisfy the hourly load with a prescribed probability. In equation (5.74), $LOLP_t$ is a reliability criterion for satisfying system reserve requirements. By properly setting $LOLP_t$ as a trade-off between economics and reliability, the chance constraint (5.74) guarantees the supply of sufficient reserves in the real-time dispatch. Constraint (5.76), formulated as a chance constraint, indicates that the base case stochastic line flow will not exceed the line capacity with a prescribed probability. Here, we assume that hourly variations of loads and renewable generation profiles follow certain continuous probability distribution functions, that is, $\tilde{D}_t$ and $\tilde{P}_{wt}$ are random variables:

$$\Pr\left\{\sum_i (P_{it} + R_{it}) + \sum_v (P_{dc,vt} - P_{c,vt} + R_{vt}) \geq \tilde{D}_t - \sum_w \tilde{P}_{wt}\right\} \geq 1 - \text{LOLP}_t \tag{5.74}$$

$$\begin{aligned} R_{it} &= I_{it} \min\left\{R_i^{\max}, P_i^{\max} - P_{it}\right\} \\ R_{vt} &= I_{vt} \min\left\{R_v^{\max}, P_{dc,v}^{\max} - P_{dc,vt}, P_{c,vt} - P_{c,v}^{\min}\right\} \end{aligned} \tag{5.75}$$

$$\Pr\left\{\left|\sum_m \text{LSF}_l^m \left(\sum_{i:i\in U(m)} P_{it} + \sum_{w:w\in U(m)} \tilde{P}_{wt}\right.\right.\right.$$
$$\left.\left.\left. + \sum_{v:v\in U(m)} (P_{dc,vt} - P_{c,vt}) - \tilde{D}_{mt}\right)\right| \leq \text{PL}_l^{\max}\right\} \geq 1 - \text{TLOP}_{l,t} \tag{5.76}$$

Considering contingency cases, the preventive (pre-contingency) and corrective (post-contingency) dispatches (5.17)–(5.23) are used for responding to contingencies by generating units and batteries. Similar to the base case, equations (5.77) and (5.78) represent the total generation and reserves which would satisfy the hourly load with a prescribed probability for contingency c, and equation (5.79) indicates that the probability of the transmission line being overloaded would be less than a preassigned value in contingency c. A proper selection of $\text{TLOP}_{l,t}$ for potential contingencies represents a balanced economic and secure operation of power systems:

$$\Pr\left\{\sum_i \left(P_{it}^c + R_{it}^c\right) + \sum_v \left(P_{dc,vt}^c - P_{c,vt}^c + R_{vt}^c\right) \geq \tilde{D}_{vt}^c - \sum_w \tilde{P}_{wt}^c\right\} \geq 1 - \text{LOLP}_t \tag{5.77}$$

$$\begin{aligned} R_{it}^c &= I_{it} \min\left\{R_i^{c,\max}, P_i^{c,\max} - P_{it}^c\right\} \\ R_{vt}^c &= I_{vt} \min\left\{R_v^{c,\max}, P_{dc,v}^{c,\max} - P_{dc,vt}^c, P_{c,vt}^c - P_{c,v}^{c,\min}\right\} \end{aligned} \tag{5.78}$$

$$\Pr\left\{\left|\sum_m \text{LSF}_l^{c,m} \left(\sum_{i:i\in U(m)} P_{it}^c + \sum_{w:w\in U(m)} \tilde{P}_{wt}^c\right.\right.\right.$$
$$\left.\left.\left. + \sum_{v:v\in U(m)} \left(P_{dc,vt}^c - P_{c,vt}^c\right) - \tilde{D}_{mt}^c\right)\right| \leq \text{PL}_l^{c,\max}\right\} \geq 1 - \text{TLOP}_{l,t} \tag{5.79}$$

Here, load and renewable generation forecast errors are represented by normal distributions with their means equal to 0 and the standard deviations proportional to the hourly forecasted values in equations (5.80) and (5.81). Thus, hourly chance constraints (5.74), (5.76), (5.77), and (5.79) can be converted into four sets of equivalent deterministic linear inequalities (5.82)–(5.85) [46, 47]. $z_{1-\text{LOLP}_t}$, $z_{1-\text{TLOP}_{l,t}/2}$, and $z_{(1-TLOP_{l,t}/2-\alpha_t^c/2)}$ are the $(100 \cdot LOLP_t)$th percentile, the $(100 \cdot TLOP_{l,t}/2)$th percentile, and the $(100 \cdot (TLOP_{l,t}/2 + \alpha_t^c/2))$th percentile of the standard normal distribution:

$$\tilde{D}_{mt} \sim D_{mt} + N\left(0, e_{mt}^2\right) \tag{5.80}$$

$$\tilde{P}_{wt} \sim P_{wt} + N\left(0, e_{wt}^2\right) \tag{5.81}$$

$$\sum_i (P_{it} + R_{it}) + \sum_v (P_{dc,vt} - P_{c,vt} + R_{vt})$$

$$\geq D_t - \sum_w P_{wt} + z_{1-LOLP_t} \left[\sum_m e_{mt}^2 + \sum_w e_{wt}^2\right]^{1/2} \tag{5.82}$$

$$PL_l^{max} \geq \left|\sum_m LSF_l^m \left(\sum_{i:i\in U(m)} P_{it} + \sum_{v:v\in U(m)} (P_{dc,vt} - P_{c,vt}) + \sum_{w:w\in U(m)} P_{wt} - D_{mt}\right)\right|$$

$$+ z_{1-TLOP_{l,t}/2} \left[\sum_m (LSF_l^m e_{mt})^2 + \sum_{w:w\in U(m)} \left(LSF_l^m e_{wt}\right)^2\right] \tag{5.83}$$

$$\sum_i \left(P_{it}^c + R_{it}^c\right) + \sum_v \left(P_{dc,vt}^c - P_{c,vt}^c + R_{vt}^c\right)$$

$$\geq D_t - \sum_w P_{wt} + z_{1-LOLP_t} \left[\sum_m e_{mt}^2 + \sum_w e_{wt}^2\right]^{1/2} \tag{5.84}$$

$$PL_l^{max} \geq \left|\left(\sum_m LSF_l^{c,m} \left(\sum_{i:i\in U(m)} P_{it}^c + \sum_{v:v\in U(m)} \left(P_{dc,vt}^c - P_{c,vt}^c\right)\right) + \sum_{w:w\in U(m)} P_{wt}^c - D_{mt}^c\right)\right|$$

$$- z_{(1-TLOP_{l,t}/2-\alpha_t^c/2)} \left[\sum_m \left(LSF_l^{c,m} e_{mt}\right)^2 + \sum_{w:w\in U(m)} \left(LSF_l^{c,m} e_{wt}\right)^2\right] \tag{5.85}$$

Suppose the operating state of each generating unit or transmission line follows a two-state continuous-time Markov chain and the initial state of system component is available. Then, α_t^c in equation (5.85) is calculated as equation (5.86), where λ_c and μ_c are failure and repair rates of the component that is out of service in contingency c:

$$\alpha_t^c = \frac{\lambda_c}{\lambda_c + \mu_c} - \frac{\lambda_c}{\lambda_c + \mu_c} e^{-(\lambda_c+\mu_c)t}, \quad \forall m, \forall t \tag{5.86}$$

The above deterministic equivalent CCO-based SCUC formulation shown in (5.1) and (5.2), (5.4)–(5.14), (5.17)–(5.23), (5.75), (5.78), and (5.82)–(5.85) is an MIP problem. Given that the day-ahead scheduling may include thousands of generating units, transmission lines, and possible contingences in a large-scale power system, the problem may not be solved without decomposition. Numerical decomposition methods discussed in Sections 5.2 and 5.4 are considered for the solution of the CCO-based SCUC problem [48].

5.4.4 Comparison of Solution Approaches

The SP-based SCUC formulation is well positioned for incorporating various uncertainties into the SCUC problem and has been widely explored in literature. In addition, although the two-stage model is made ahead of real-time operation, it could consider the anticipated reactions of the system operator when more information becomes available in real time through the second-stage scenario-dependent formulation. However, one key issue in the SP-based SCUC is to generate scenarios that would truly reflect probabilistic characteristics of various power system uncertainties. The scenario-based approach acknowledges a given probability distribution for representing system uncertainties, and it is usually difficult to justify if they could truly represent actual uncertainty factors, which may impact the quality of the SP-based SCUC solution. In addition, the quality of solutions could critically depend on the choice of the scenario set in the sense that having more scenarios usually leads to a more accurate solution. However, a larger set of scenarios also means a significant number of variables and constraints, which would significantly increase the size of the two-stage SP-based SCUC problem and be computationally expensive.

The RO-based SCUC approach does not require an explicit knowledge of probability distributions. Alternatively, it adopts deterministic uncertainty sets for representing various uncertainties, which can be constructed using a limited set of information about the uncertainty, namely the mean value and some estimates of the variance or a range of possible variations around the mean value. The RO-based SCUC approach would protect the system against every possible realization of uncertainties contained in the chosen set. However, solution conservatism is one major concern of the RO-based SCUC approach. Uncertainty intervals would need to be carefully selected and the budget level would need to be carefully tuned. A narrow confidence interval and/or a smaller budget level may not cover the entire uncertainty spectrum and, in turn, lead to a UC solution that would not correspond to all possible uncertain situations. On the other hand, a wide interval and/or a higher budget level could lead to pessimistic solutions which would not utilize system resources efficiently and be of limited use to system operators. In addition, the RO-based approach would solve max–min subproblems which could be computationally intractable, especially when the inner minimization problem is a non-linear programming (NLP) problem (considering the AC power flow) or an MIP problem (considering UC adjustments of quick-start units).

The CCO-based SCUC approach may provide more robust solutions than those of the SP-based SCUC, because the CCO-based SCUC solution is independent of the choice of scenarios. On the other hand, it suffers from similar restrictions as those of the SP-based SCUC, which require an explicit knowledge of the probability distribution of uncertainties. Furthermore, CCO-based SCUC models are usually more difficult to solve than the SP-based SCUC models. There is a closed link between the RO-based SCUC and the CCO-based SCUC. For instance, one can design an uncertainty set in such a way that it would enforce a certain probabilistic constraint, so that the RO-based SCUC solutions would be comparable to those derived from the CCO-based SCUC.

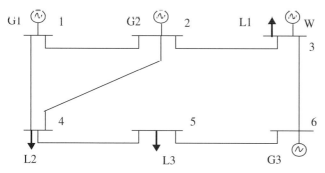

Figure 5.8   One-line diagram of the six-bus system.

The three SCUC formulations with uncertainties may represent different performances, including solution robustness in the face of uncertainties, the cost of generation schedules, and the computational cost. The proposed techniques may be effectively combined for increasing the quality of the SCUC solution and/or enhancing the computational performance of the proposed solution algorithms. For instance, SP and CCO are applied together in Reference [45] and a unified SP and RO UC model is discussed in Reference [49].

5.5 ILLUSTRATIVE RESULTS

The six-bus system shown in Figure 5.8 is used to illustrate the performance of the SP-based SCUC, the RO-based SCUC, and the CCO-based SCUC approaches. The numerical test results of the three approaches on larger systems can be found in the authors' previous work [40, 50, 51]. The illustrated system has three thermal generators, one wind farm, three loads, and seven transmission lines [51, 52]. Generator data are shown in Table 5.4, transmission line data are given in Table 5.5, and mean up/down times of generators and transmission lines are presented in Table 5.6. G1 is a cheap coal-fired unit, while G2 and G3 are more expensive oil-fired and gas-fired units, respectively. All three generators are modeled as non-quick-start units in the

TABLE 5.4   Generator data

U	Bus no.	a ($/h)	b ($/MWh)	c ($/MW^{2}h)	Pmax (MW)	Pmin (MW)	Ini. st. (h)	Start up (MBtu)	Corrective capacity (MW)
		Unit cost coefficients							
G1	1	176.9	13.5	0.0004	220	100	4	100	9.16
G2	2	129.9	32.6	0.0010	100	10	2	200	8.33
G3	6	137.4	17.6	0.0050	20	10	1	0	6.49

TABLE 5.5 Transmission line data

Line	From	To	X(pu)	Flow limit (MW)
L1	1	2	0.170	200
L2	2	3	0.037	100
L3	1	4	0.258	100
L4	2	4	0.197	100
L5	4	5	0.037	100
L6	5	6	0.140	100
L7	3	6	0.018	100

TABLE 5.6 Mean up/down time of generators and lines (h)

	Mean up time	Mean down time		Mean up time	Mean down time
G1	23.6	0.4	L3	23.6	0.4
G2	23.7	0.3	L4	23.6	0.4
G3	23.8	0.2	L5	23.7	0.3
L1	23.5	0.5	L6	23.6	0.4
L2	23.7	0.3	L7	23.8	0.2

example, that is, the UC status of all three generators in the base case and under uncertainties are the same. The impact of quick-start units on the three approaches when considering larger systems can be found in the authors' previous work [38, 53, 54].

The system is tested for a two-hour case with forecasted loads of 175.19 MW and 265.19 MW, and forecasted wind generation of 58.15 MW and 18.39 MW. The following four cases are studied here:

Case 1. Deterministic SCUC study.

Case 2. SP-based SCUC study (load and wind uncertainties are not correlated).

Case 3. RO-based SCUC study.

Case 4. CCO-based SCUC study.

In all four cases, the impact of grid-scale batteries on the operational performance of power systems is also explored by adding a grid-connected battery at bus 3. Parameters of the battery are shown in Table 5.7.

TABLE 5.7 Battery data

Emax (MWh)	Emin (MWh)	E initial (MWh)	E terminal (MWh)	Maximum charging/discharging (MW)	Minimum charging/discharging (MW)	Efficiency
20	10	15	15	10	2	90%

TABLE 5.8   Deterministic SCUC result without battery

Hour	G1 (MW)	G2 (MW)	G3 (MW)	Wind (MW)	Total cost ($)
1	137.59	0	0	37.60	7278.1
2	191.59	35.21	20.00	18.39	

Case 1: Deterministic SCUC study

The deterministic SCUC results without a battery are shown in Table 5.8. When load and wind forecasting inaccuracies are not considered, large and economic coal-burning unit G1 is dispatched as extensively to supply the system load while the expensive G2 and G3 are dispatched as required to balance the system load. In addition, by enforcing the transmission capacity, available wind generation of 58.15 MW at hour 1 is not fully utilized. When a battery is connected to bus 3, the results in Table 5.9 show that more wind energy can be utilized at hour 1, which is used to charge the battery and discharge it at hour 2 for supplying the peak load. Accordingly, the total cost is lower than that in Table 5.8. In Table 5.9, negative values represent the amount of energy supplied to the battery.

Case 2: SP-based SCUC study, in which load and wind uncertainties are assumed to follow independent normal distributions

In this case, load and wind uncertainties are assumed to follow independent normal distributions $N(175.19, 7.15)$ and $N(58.15, 5.93)$ at hour 1, and $N(265.19, 10.82)$ and $N(18.39, 1.88)$ at hour 2. The standard deviations are determined so that the 3σ range would cover load uncertainty range of $[175.19 - 0.08 \times 175.19, 175.19 + 0.08 \times 175.19]$ and $[265.19 - 0.08 \times 265.19, 265.19 + 0.08 \times 265.19]$ and wind uncertainty range of $[58.15 - 0.2 \times 58.15, 58.15 + 0.2 \times 58.15]$ and $[18.39 - 0.2 \times 18.39, 18.39 + 0.2 \times 18.39]$, which will be used in Case 3 for the RO-based SCUC study, that is, $\sigma_{load,1} = 0.08 \times 175.19/1.96 = 7.15$, $\sigma_{load,2} = 0.08 \times 265.19/1.96 = 10.82$, $\sigma_{wind,1} = 0.20 \times 58.15/1.96 = 5.93$, and $\sigma_{wind,2} = 0.20 \times 18.39/1.96 = 1.88$. Table 5.10 shows the results of SP-based SCUC with two independent scenario tests. For both tests, the expected dispatch of G1 is smaller than that of the deterministic result, and the expected dispatch of G2 is higher. The reason is that in order to accommodate various uncertainties, the SP-based SCUC results would dispatch more expensive generators and, in turn, the expected operation cost is higher than that of the deterministic case in Table 5.8. In addition, in

TABLE 5.9   Deterministic SCUC result with battery

Hour	G1 (MW)	G2 (MW)	G3 (MW)	Wind (MW)	Battery (MW)	Total cost ($)
1	138.09	0.00	0	42.10	−5.56	7113.2
2	192.09	29.71	20.00	18.39	5.00	

TABLE 5.10   SP-based SCUC results without battery

Test		1		2	
# of scenarios		100		1000	
Hour		1	2	1	2
G1 (MW)	Expected dispatch	134.23	188.23	134.95	188.95
	Std of dispatch	6.11	6.11	5.40	5.40
G2 (MW)	Expected dispatch	10.34	39.07	10.20	38.12
	Std of dispatch	1.73	7.68	1.45	6.46
G3 (MW)	Expected dispatch	0	19.64	0	19.80
	Std of dispatch	0	1.38	0	1.03
W (MW)	Expected dispatch	30.62	18.26	30.04	18.32
	Std of dispatch	5.26	0.72	4.84	0.66
Cost ($)	Expected cost	7365.0		7353.7	
	Std of cost	69.53		64.73	

Table 5.10, the tests with different number of original scenarios have close expected dispatches for all generators as well as the expected operation cost. Furthermore, standard deviations of generation dispatches and operation costs for scenarios in test 2 are smaller than those of test 1, which shows that the SP-based SCUC approach would derive more stable results when a larger number of scenarios is considered. Table 5.11 shows the SP-based SCUC results with battery. The similar observation is obtained as the deterministic case, which indicates that the inclusion of battery

TABLE 5.11   SP-based SCUC results with battery

Test		1		2	
# of scenarios		100		1000	
Hour		1	2	1	2
G1 (MW)	Expected dispatch	136.38	190.38	136.54	190.54
	Std of dispatch	4.52	4.52	4.22	4.22
G2 (MW)	Expected dispatch	10.31	31.53	10.22	31.33
	Std of dispatch	1.85	5.02	1.54	4.56
G3 (MW)	Expected dispatch	0.00	19.91	0.00	19.94
	Std of dispatch	0.00	0.61	0.00	0.49
W (MW)	Expected dispatch	43.50	18.38	43.44	18.38
	Std of dispatch	3.48	0.24	3.47	0.23
Battery (MW)	Expected dispatch	−5.56	0.00	−5.56	
	Std of dispatch	0.00	5.00	0.00	5.00
Cost ($)	Expected cost	7122.6		7120.8	
	Std of cost	41.86		35.39	

would enhance the utilization of available wind generation and reduce the expected operation cost. In addition, the expected cost of the SP-based SCUC with battery is close to that of the deterministic case, which indicates that including batteries would help diminish the impacts of uncertainties on power system operation, in terms of the total operation cost.

Case 3: RO-based SCUC study

In this case, uncertainty intervals are derived by load and wind forecast values D_{mt} and $P_{f,wt}$ together with the load variation of $0.08D_{mt}$ and the wind variation of $0.2P_{f,wt}$, that is, the load uncertainty intervals are $[175.19 - 0.08 \times 175.19, 175.19 + 0.08 \times 175.19]$ and $[265.19 - 0.08 \times 265.19, 265.19 + 0.08 \times 265.19]$ and the wind uncertainty intervals are $[58.15 - 0.2 \times 58.15, 58.15 + 0.2 \times 58.15]$ and $[18.39 - 0.2 \times 18.39, 18.39 + 0.2 \times 18.39]$ for the two hours; $0.08D_{mt}$ and $0.2P_{f,wt}$ are system specific. A narrow uncertainty interval may not cover the entire uncertainty spectrum and, in turn, lead to a SCUC solution that would not correspond to all possible uncertain situations. On the other hand, a wide uncertainty interval could lead to pessimistic solutions which would not utilize system resources efficiently and be of limited use to system operators. Based on authors' experience, $0.08D_{mt}$ and $0.2P_{f,wt}$ could cover about 85% of NYISO historical load and wind realizations in June–August 2013 [50].

Table 5.12 reports the RO-based SCUC results with respect to two load and wind uncertainty budget levels. Besides generation dispatch and costs for the base case, Table 5.12 also shows the results with respect to the best and the worst load and wind realizations, in terms of the lowest and the highest operation costs. Thus, the RO-based SCUC approach also provides an operation cost interval. It shows that when the budget level increases, UC costs are monotonously increasing because wind

TABLE 5.12   RO-based SCUC result without battery

Budget level	Case/hour		G1 (MW)	G2 (MW)	G3 (MW)	Wind (MW)	Load (MW)	Cost ($)
0.5	Base	1	128.18	10.00	0.00	37.00	175.19	7363.3
		2	182.18	44.61	20.00	18.39	265.19	
	Best	1	130.56	18.33	0.00	19.29	168.18	7311.3
		2	184.56	36.28	13.51	20.23	254.58	
	Worst	1	132.30	18.33	0.00	31.56	182.20	8029.1
		2	186.30	52.94	20.00	16.55	275.80	
1.0	Base	1	117.85	14.35	0.00	42.99	175.19	7658.2
		2	171.85	62.35	13.51	17.48	265.19	
	Best	1	108.69	10.00	0.00	42.48	161.17	7099.4
		2	162.69	54.02	20.00	7.26	243.97	
	Worst	1	127.01	22.68	0.00	39.51	189.21	8595.5
		2	181.01	70.68	20.00	14.71	286.41	

TABLE 5.13   RO-based SCUC result with battery

Budget level	Case/ hour		G1 (MW)	G2 (MW)	G3 (MW)	Battery (MW)	Wind (MW)	Load (MW)	Cost ($)
0.5	Base	1	134.18	0.00	0.00	−5.56	46.56	175.19	7125.3
		2	188.18	33.62	20.00	5.00	18.39	265.19	(3.23%)
	Best	1	131.56	0.00	0.00	2.00	38.62	168.18	7031.6
		2	185.56	25.28	13.51	10.00	20.23	254.58	(3.83%)
	Worst	1	143.34	0.00	0.00	2.00	40.86	182.20	7346.8
		2	187.30	41.95	19.99	10.00	16.55	275.80	(8.50%)
1.0	Base	1	118.85	10.00	0.00	5.56	51.89	175.19	7262.3
		2	172.85	51.35	17.59	5.00	18.39	265.19	(5.17%)
	Best	1	109.69	10.00	0.00	2.00	43.48	161.17	6870.5
		2	163.69	43.02	13.19	2.00	22.07	243.97	(3.22%)
	Worst	1	128.01	16.67	0.00	2.00	46.52	189.21	7593.8
		2	182.01	59.68	20.00	10.00	14.71	286.41	(11.65%)

utilization is decreased and expensive units need to be committed more for covering a wider range of uncertainties.

When the battery is included, the similar results on enhancing the utilization of available wind generation and reducing the operation cost are observed in Table 5.13. In particular, with the budget level of 1.0, the worst-case cost with battery (i.e., $7593.8) is even better than the base case cost without battery (i.e., $7658.2). The last column in Table 5.13 shows the cost reduction, which is calculated as (the cost in Table 5.12 – the cost in Table 5.13)/the cost in Table 5.13. More significant cost improvements are observed with higher budget levels. In particular, the worst-case operation cost with the uncertainty budget level of 1.0 has been improved by 11.65% with batteries. This indicates that batteries are particularly important for highly variable power systems when a budget level with higher uncertainty is considered.

Case 4: CCO-based SCUC study

In this case, power system uncertainties are simulated via probability constraints. Table 5.14 shows the results with six different sets of LOLP and TLOP. The power system operation cost is decreased as TLOP is increased for a fixed LOLP, which is because line flow constraints become less stringent. Similarly, for a fixed TLOP, the system operation cost is decreased as LOLP is increased, which is because a smaller system reserve will be required. The minimum operation cost of $7695.3 is obtained when LOLP = 0.1 and TLOP = 0.4, while the maximum operation cost of $8625.2 occurs when LOLP = 0.05 and TLOP = 0.05. It is also noted that the minimum operation cost of $7695.3 is close to that of the RO-based SCUC of $7658.2 in Table 5.12, which indicates that one may get similar performances for RO-based SCUC and CCO-based SCUC when parameters are properly adjusted. In addition, similar performances on enhancing the utilization of available wind power

TABLE 5.14   CCO-based SCUC result without battery

LOLP/ TLOP	Hour	G1 (MW)	G2 (MW)	G3 (MW)	Wind (MW)	Total cost ($)
0.05/0.05	1	126.12	31.18	0	17.88	8625.2
	2	151.92	79.18	15.70	18.39	
0.05/0.2	1	139.63	17.67	0	17.88	8136.8
	2	165.43	65.67	15.70	18.39	
0.05/0.4	1	147.31	10	0	17.88	7842.3
	2	174.19	56.91	15.70	18.39	
0.1/0.05	1	125.85	26.24	0	23.10	8387.9
	2	152.56	74.24	20	18.39	
0.1/0.2	1	139.36	12.73	0	23.10	7899.9
	2	166.07	60.73	20	18.39	
0.1/0.4	1	142.09	10	0	23.10	7695.3
	2	174.83	51.97	20	18.39	

generation and reducing the operation cost are observed as we consider a battery, as shown in Table 5.15. In addition, the last column in Table 5.15 shows the operation cost reduction, which is calculated as (the cost in Table 5.14 – the cost in Table 5.15)/the cost in Table 5.14. More significant cost improvements are observed with smaller values of LOLP and TLOP, which indicates that batteries are particularly important for highly variable systems.

When comparing the operation costs derived by the four methods without battery as shown in Tables 5.7, 5.9, 5.11, and 5.13, the deterministic SCUC derives the

TABLE 5.15   CCO-based SCUC result with battery

LOLP/ TLOP	Hour	G1 (MW)	G2 (MW)	G3 (MW)	Battery (MW)	Wind (MW)	Total cost ($)
0.05/0.05	1	101.85	20.74	0	−5.56	58.15	7686.4
	2	153.06	68.74	20	5	18.39	(10.88%)
0.05/0.2	1	112.60	10	0	−5.56	58.15	7248.8
	2	166.57	55.23	20	5	18.39	(10.91%)
0.05/0.4	1	121.33	10	0	−5.56	49.41	7221.2
	2	175.33	46.47	20	5	18.39	(7.92%)
0.1/0.05	1	101.86	20.74	0	−5.56	58.15	7686.4
	2	153.06	68.74	20	5	18.39	(8.36%)
0.1/0.2	1	112.60	10	0	−5.56	58.15	7248.8
	2	166.57	55.23	20	5	18.39	(8.24%)
0.1/0.4	1	122.60	0	0	−5.56	58.15	7183.6
	2	175.33	46.47	20	5	18.39	(6.65%)

TABLE 5.16 Robustness test of the four models—average load shedding

Case		Without battery (MWh)	With battery (MWh)	Improvement
Deterministic SCUC		1.7698	0.6558	62.94%
SP-based SCUC	Test 1	1.0566	0.4896	53.66%
	Test 2	1.0001	0.4345	56.55%
RO-based SCUC	0.5 budget	0.6260	0.4600	26.52%
	1.0 budget	0.0761	0.0717	5.78%
CCO-based SCUC	0.05/0.05	0.1079	0.1012	6.21%
	0.05/0.2	0.1147	0.1102	3.92%
	0.05/0.4	0.1642	0.1437	12.48%
	0.1/0.05	0.2670	0.1279	52.09%
	0.1/0.2	0.2985	0.1402	53.03%
	0.1/0.4	0.3990	0.2282	42.81%

cost of $7278.1. The SP-based SCUC provides the operation cost of $7365.0 ± 136.26 for MCS with a 95% confidence interval in test 1, that is, 95% of the operation cost will be within the given interval of [$7228.74, $7501.26]. The RO-based SCUC shows the operation cost interval of [$7099.4, $8595.5] for 1.0 uncertainty budget level. In comparison, the cost derived by the deterministic SCUC is within the intervals provided by SP-based and RO-based SCUC cases. In addition, the RO-based SCUC provides a much wider interval than that of the SP-based SCUC. This is particularly the case when a wider range of uncertainties is considered. In addition, comparing results in Tables 5.9 and 5.11, it shows that the RO-based SCUC solution is very sensitive to the budget level and the SP-based SCUC solution is relatively stable considering various tests. Furthermore, most of the CCO-based SCUC results are within the interval [$7099.4, $8595.5] offered by the RO-based SCUC results, except the one with LOLP = 0.05 and TLOP = 0.05. This again shows that with a proper choice of LOLP and TOLP, one can find the equivalent results for CCO- and RO-based SCUC cases.

The robustness performance of the four models is tested by measuring the expected load shedding of a new independent set of 100 scenarios. The robustness test is performed by solving the redispatch problem for each scenario, which is restricted by the UC and dispatch solutions in the base case as well as corrective capacities of thermal generators. The value of lost load (VOLL) is set at 10k$/MWh for penalizing possible load sheddings in scenarios. Table 5.16 reports the average load shedding for the four models. It shows that the deterministic SCUC would encounter significantly higher load shedding as compared to the other three models, which indicates that sophisticated models (beyond deterministic approaches) are necessary in order to accommodate uncertainties in SCUC solutions and enhance the security of power systems. For the RO-based SCUC model, a higher budget level would induce a smaller load shedding as more uncertainties will be covered. Similarly, for the SP-based SCUC, a larger number of scenarios would induce a smaller load shedding as additional scenarios derive a better SCUC solution for handling uncertainties. For the CCO-based SCUC, smaller LOLP and TLOP would lead to a smaller load shedding as the power system will be operated more comprehensively to protect it against stringent situations. In addition, if a proper uncertainty budget level and LOLP/TLOP are set, the performances of RO-based SCUC and CCO-based SCUC would be comparable and better than that of the SP-based SCUC in terms of the load shedding quantity.

The last column in Table 5.16 lists the improvement in load shedding, which is calculated as (load shedding without battery – load shedding with battery)/load shedding without battery. Solutions that have a worse robustness performance in terms of larger load shedding quantities would have a higher load shedding improvement when the battery is used. The deterministic SCUC case has the highest improvement of 62.94% which indicates that the battery is particularly important to power system operations with uncertainties, especially when the comprehensively robust operation decisions are hard to obtain.

The applications of the three SCUC formulations with uncertainties to larger power systems were explored in References [48, 50, 51, 53] in which the comparative computation performances indicate that they are suitable alternatives in practical power systems for solving the day-ahead scheduling problem with uncertainties. The main obstacle in solving the SP-based SCUC is the number of scenarios. For larger power systems, as long as the deterministic SCUC can be effectively solved, the proposed BD approach could handle a large number of scenarios in subproblems and send a feedback as valid Benders cuts to the master problem, which represent a similar problem to that of the deterministic SCUC. The major computation complexity of the RO-based SCUC is in solving the max–min subproblem, especially when the inner minimization problem is NLP (when considering the AC power flow) or MIP (when considering UC adjustment of quick-start units). The future work on this subject can explore alternative and more efficient techniques to further improve the computation performance of the RO-based SCUC approach. For instance, instead of solving the mixed-integer problem with complementarity constraints, we may iteratively solve LP problems with complementarity constraints and restrict integer variables in an outer loop. The future work can also explore a column-and-constraint generation algorithm and those for accelerating BD to further strengthen the cutting planes and accelerate the convergence performance of the modified BD method. Depending on the solution approach, the problem scale and the computation burden of the CCO-based SCUC models for large power systems are comparable to those of scenario-based and interval optimization approaches.

5.6 CONCLUSIONS

This chapter reviews the state-of-the-art approaches for solving deterministic SCUC and SCUC with uncertainties (including SP-based SCUC, RO-based SCUC, and CCO-based SCUC models). Although SCUC has been extensively studied by power academic and industrial societies, it still cannot be presently considered as a well-solved problem, especially in observing various uncertainties in emerging power systems which bring additional dimensions of complexity to SCUC. Research on such an extremely challenging problem is of great importance for improving the efficiency and security of electrical power systems. Thus, there are plenty of opportunities to make contributions to modeling and solving this challenging class of SCUC problems:

- Improving the computational performance of the deterministic SCUC. Tighter MIP formulation of the deterministic SCUC may decrease the number of nodes of the search tree via cutting down the number of integer variables in the branch-and-cut method. It is also possible to add a proper number of redundant constraints for driving the LP solution toward integer solution and user-defined cutting planes for shrinking the feasible region. In addition, usually most security constraints in SCUC are inactive or redundant especially during the off-peak load periods, and such inactive security constraints do not affect the final SCUC solution. Consequently, the SCUC problem scale will be greatly reduced, and a faster SCUC solution can be obtained.

- Combining SP, RO, and CCO in an effective way for modeling and solving SCUC with uncertainties, in order to increase the quality of the solution and enhance the computational performance.

- Decomposition approaches for deterministic SCUC and SCUC with uncertainties provide the possibility of using the *master/slave* parallel computing *paradigm*, by solving subproblems with each slave, and the master problem with the master. The parallel computing paradigm needs to be carefully designed for balancing the size and the computational burden of each subproblem on each processor, as well as the time for coordinating tasks on different processors.

APPENDIX 5.A NOMENCLATURE

a, b, m	Indices of buses
c, s	Indices of contingences and scenarios
c_{ik}	Incremental cost for segment k of unit i
D_t, R_{St}, R_{Ot}	System load demand, spinning reserve requirement, and operating reserve requirement at hour t
D_{mt}	Total load of bus m at hour t
$D_{mt}^u, P_{f,wt}^u$	Sets of possible load and wind realizations
DP_i, UP_i	Shutdown/start-up ramp limits of unit i
DR_i, UR_i	Ramping down and ramping up limits of unit i
DT_i, UT_i	Numbers of hours unit i must be initially offline and online due to its minimum off/on time limits

E_{vt}	Energy stored in battery v at hour t
$E_v^{\min}$, $E_v^{\max}$	Energy lower and upper limits of battery v at hour t
E_{v0}, E_{vNT}	Initial and terminal energy levels of battery v at hour t
i, t, k, l	Indices for units, hours, segments of cost curve functions, and lines
I_{it}	Unit commitment decision of unit i at hour t
$I_{dc,vt}, I_{c,vt}$	Charging and discharging indicators of battery v at hour t
LOLP_t	Loss-of-load probability at hour t
LSF_l^m	The sensitivity of power flow of line l to power injection at bus m
MSR_i, QSC_i	Spinning reserve and quick-start capacity that can be provided by unit i in 1 min
N_i	No-load cost of unit i
$\text{OR}_{it}, \text{SR}_{it}$	Operating and spinning reserve provided by unit i at hour t
OR_{vt}, SR_{vt}	Operating and spinning reserve provided by battery v at hour t
p^s	Probability of scenario s
$P_{f,wt}$	Forecasted available wind generation
$P_{dc,vt}, P_{c,vt}$	Power discharged and charged by battery v at hour t
$P_{c,v}^{\min}, P_{c,v}^{\max}$	Charging power lower and upper limits of battery v at hour t
$P_{dc,v}^{\min}, P_{dc,v}^{\max}$	Discharging power lower and upper limits of battery v at hour t
PL_{lt}	Power flow of transmission line l at hour t
$\text{PL}_l^{\max}$	Power flow upper limit of transmission line l
$P_{dc,vt}, P_{c,vt}$	Power discharged and charged by battery v at hour t
P_{ikt}	Generation dispatch of unit i at hour t at segment k
$P_{ik}^{\max}$	Maximum capacity of unit i of segment k
$P_i^{\min}, P_i^{\max}$	Minimum/maximum capacity of unit i
P_{it}	Generation dispatch of unit i at hour t
P_{wt}	Generation dispatch of wind farm w at hour t
$P_{it}^u(D_{mt}^u, P_{f,wt}^u)$	Sets of variables representing all possible dispatches of thermal units that could be adaptively and securely adjusted in response to all possible load and wind generation realizations
$P_{wt}^u(D_{mt}^u, P_{f,wt}^u)$	Sets of variables representing all possible dispatches of wind farms that could be adaptively and securely adjusted in response to all possible load and wind generation realizations
$\text{SU}_{it}, \text{SD}_{it}$	Start-up and shut down costs of unit i at hour t
$T_i^{\text{on}}, T_i^{\text{off}}$	Minimum on/off time limits of unit i
$\text{TLOP}_{l,t}$	Probability of transmission line l overload at hour t
$U(m)$	Set of generators/wind farms/batteries located at bus m

$X_{i0}^{on}, X_{i0}^{off}$ On/off time counter of unit i at the initial status

λ, μ Dual variables

η_v Efficiency of battery v

Δ_{it}^c Corrective capacity of unit i at hour t in contingency c

Δ_d and Δ_w Budget limits for load and wind generation uncertainties

ACKNOWLEDGMENTS

This report was supported in part by the US National Science Foundation grants ECCS-1102064 and ECCS-1254310 and the US Department of Energy grants DE-FOA-0000856 and DE-EE-0001380.

REFERENCES

[1] K. Abdul-Rahman, M. Shahidehpour, M. Aganagic, and S. Mokhtari, "A practical resource scheduling with OPF constraints," *IEEE Transactions on Power Systems*, vol. 11, no. 1, pp. 254–259, February 1996.

[2] C. Wang and M. Shahidehpour, "Ramp-rate limits in unit commitment and economic dispatch incorporating rotor fatigue effect," *IEEE Transactions on Power Systems*, vol. 9, no. 3, pp. 1539–1545, August 1994.

[3] S. Wang, M. Shahidehpour, D. Kirschen, S. Mokhtari, and G. Irisarri, "Short-term generation scheduling with transmission and environmental constraints using an augmented Lagrangian relaxation," *IEEE Transactions on Power Systems*, vol. 10, no. 3, pp. 1294–1301, August 1995.

[4] N. Balu, T. Bertram, A. Bose, V. Brandwajn, G. Cauley, D. Curtice, A. Fouad, L. Fink, M. Lauby, B. Wollenberg, and J. Wrubel, "On-line power system security analysis," *Proceedings of the IEEE*, vol. 80, no. 2, pp. 262–280, February 1992.

[5] North American Electric Reliability Council, *Reliability Concepts in Bulk Power Electric Systems*, 1985.

[6] B. Stott, O. Alsac, and A. Monticelli, "Security analysis and optimization," *Proceedings of the IEEE*, vol. 75, no. 12, pp. 1623–1644, December 1987.

[7] Y. Fu, M. Shahidehpour, and Z. Li, "Security-constrained unit commitment with AC constraints," *IEEE Transactions on Power Systems*, vol. 20, no. 3, pp. 1538–1550, August 2005.

[8] M. Shahidehpour and M. Marwali, *Maintenance Scheduling in Restructured Power Systems*, Kluwer Academic Publishers, Norwell, MA, 2000.

[9] M. Shahidehpour, H. Yamin, and Z. Li, *Market Operations in Electric Power Systems*, John Wiley & Sons, Inc., New York, 2002.

[10] U. S. Department of Energy, "The smart grid: an introduction," 2008. Available online at http://www.oe.energy.gov/DocumentsandMedia/DOE_SG_Book_Single_Pages(1).pdf.

[11] A. J. Wood and B. F. Wollenberg, *Power Generation, Operation and Control*, John Wiley & Sons, NY, 1984.

[12] L. Wu, "A tighter piecewise linear approximation of quadratic cost curves for unit commitment problems," *IEEE Transactions on Power Systems*, vol. 26, no. 4, pp. 2581–2583, November 2011.

[13] T. Li and M. Shahidehpour, "Price-based unit commitment: a case of Lagrangian relaxation versus mixed integer programming," *IEEE Transactions on Power System*, vol. 20, pp. 2015–2025, 2005.

[14] Y. Fu, M. Shahidehpour, and Z. Li, "AC contingency dispatch based on security-constraints unit commitment," *IEEE Transactions on Power Systems*, vol. 21, no. 2, pp. 897–908, May 2006.

[15] J. Ostrowski, M.F. Anjos, and A. Vannelli, "Tight mixed integer linear programming formulations for the unit commitment problem," *IEEE Transactions on Power Systems*, vol. 27, no. 1, pp. 39–46, February 2012.

[16] Y. Fu, Z. Li, and L. Wu, "Modeling and solution of the large-scale security-constrained unit commitment," *IEEE Transactions on Power Systems*, vol. 28, no. 4, pp. 3524–3533, November 2013.

[17] L. Wu and M. Shahidehpour "Accelerating the Benders decomposition for network-constrained unit commitment problems," *Energy Systems*, vol. 1, no. 3, pp. 339–376, August 2010.

[18] K. Brokish and J. Kirtley, "Pitfalls of Modeling Wind Power Using Markov Chains," Proceedings 2009 IEEE Power and Energy Society General Meeting, pp. 1–5, 2009.

[19] Y. Li, K. Xie, and B. Hu, "Copula-ARMA model for multivariate wind speed and its application in reliability assessment of generating systems," *IEEE Transactions on Energy Conversion*, vol. 8, pp. 1–7, 2013.

[20] J. Apt, "The spectrum of power from wind turbines," *Journal of Power Sources*, vol. 169, no. 2, pp. 369–374, 2007.

[21] P. Doody and S. Santoso, "A Comparative Metric to Quantify the Variability of Wind Power," Proceedings IEEE Power Energy Society General Meeting, pp. 1–6, 2009.

[22] W. Katzenstein, E. Fertig, and J. Apt, "The variability of interconnected wind plants," *Energy Policy*, vol. 38, no. 8, pp. 4400–4410, 2010.

[23] U.S. DoE Energy Information Administration, Annual Energy Outlook 2015.

[24] M. Bolinger, R. Wiser, and W. Golove, "Accounting for fuel price risk when comparing renewable to gas-fired generation: the role of forward natural gas prices," Lawrence Berkeley National Laboratory. Available at http://eetd.lbl.gov/sites/all/files/publications/report-lbnl-54751.pdf

[25] A. Alabdulwahab, A. Abusorrah, Z. Zhang, and M. Shahidehpour, "Coordination of interdependent natural gas and electricity infrastructures for firming the variability of wind energy in stochastic day-ahead scheduling," *IEEE Transactions on Sustainable Energy*, vol. 6, no. 2, pp. 606–615, May 2015.

[26] B. Bush, T. Jenkin, D. Lipowicz, and D. J. Arent, *Variance Analysis of Wind and Natural Gas Generation under Different Market Structures: Some Observations*, National Renewable Energy Laboratory, 2012.

[27] N. Keyaerts, Y. Rombauts, E. Delarue, and W. D'haeseleer, "Impact of Wind Power on Natural Gas Market: Inter Market Flexibility," Proceedings Energy Market, 7th International Conf on the European, 2010.

[28] A. Lee, O. Zinaman, and J. Logan, *Opportunities for Synergy Between Natural Gas and Renewable Energy in the Electric Power and Transportation Sectors*, National Renewable Energy Laboratory, 2012.

[29] J. Birge and F. Louveaus, *Introduction to Stochastic Programming*, 1997.

[30] A. Ben-Tal, L. Ghaoui, and A. Nemirovski, *Robust Optimization*, Princeton University Press, 2009.

[31] J. Dupacova, G. Consigli, and S.W. Wallace, "Scenarios for multistage stochastic programs," *Annals of Operations Research*, vol. 100, pp. 25–53, 2000.

[32] C. Liu, M. Shahidehpour, and L. Wu, "Extended Benders decomposition for two-stage SCUC," *IEEE Transactions on Power Systems*, vol. 25, no. 2, pp. 1192–1194, May 2010.

[33] J. F. Benders, "Partitioning procedures for solving mixed-variables programming problems," *Numerische Mathematik*, vol. 4, no. 1, pp. 238–252, 1962.

[34] N. Alguacil and A. J. Conejo, "Multiperiod optimal power flow using Benders decomposition," *IEEE Transactions on Power Systems*, vol. 15, no. 1, pp. 196–201, 2000.

[35] J. Martínez-Crespo, J. Usaola, and J.F. Fernández, "Security-constrained optimal generation scheduling in large-scale power systems," *IEEE Transactions on Power Systems*, vol. 21, no. 1, pp. 321–332, 2006.

[36] M. Shahidehpour and Y. Fu, "Benders decomposition: applying Benders decomposition to power systems," *IEEE Power and Energy Magazine*, vol. 3, no. 2, pp. 20–21, 2005.

[37] T. Li, M. Shahidehpour, and Z. Li, "Risk-constrained bidding strategy with stochastic unit commitment," *IEEE Transactions on Power Systems*, vol. 22, no. 1, pp. 449–458, 2007.

[38] L. Wu, M. Shahidehpour, and C. Liu, "MIP-based post-contingency corrective action with quick-start units," *IEEE Transactions on Power Systems*, vol. 24, no. 4, pp. 1898–1899, November 2009.

[39] L. Wu, "An improved decomposition framework for accelerating LSF and BD based methods for network-constrained UC problems," *IEEE Transactions on Power Systems*, vol. 28, no. 4, pp. 3977–3986, November 2013.

[40] H. Wu and M. Shahidehpour, "Stochastic SCUC with variable wind penetration using constrained ordinal optimization," *IEEE Transactions on Sustainable Energy*, vol. 5, no. 2, pp. 379–388, April 2014.

[41] H. Wu, M. Shahidehpour, A. Alabdulwahab, and A. Abusorrah, "Thermal generation flexibility with ramping costs and hourly demand response in stochastic security-constrained scheduling of variable energy sources," *IEEE Transactions on Power Systems*, vol. 30, no. 6, pp. 2955–2964, 2015.

[42] R. H. Byrd, M. E. Hribar, and J. Nocedal, "An interior point algorithm for large-scale nonlinear programming," *SIAM Journal of Optimization*, vol. 94, no. 4, pp. 877–900, September 1999.

[43] D. Bertsimas, E. Litvinov, X. A. Sun, J. Zhao, and T. Zheng, "Adaptive robust optimization for the security constrained unit commitment problem," *IEEE Transactions on Power Systems*, vol. 28, no. 1, pp. 52–63, February 2013.

[44] T. Szantai, "A computer code for solution of probabilistic-constrained stochastic programming problems," in *Numerical Techniques for Stochastic Optimization*, Y. Ermoliev and R.J.-B. Wets, Eds., ch. 10, pp. 230–235, Springer-Verlag, Berlin, Germany, 1988.

[45] Q. Wang, Y. Guan, and J. Wang, "A chance-constrained two-stage stochastic program for unit commitment with uncertain wind power output," *IEEE Transactions on Power Systems*, vol. 27, no. 1, pp. 206–215, February 2012.

[46] R. Henrion, "Introduction to Chance-Constrained Programming," Tutorial paper for the Stochastic Programming Community Home Page, 2004.

[47] A. Prekopa, *Stochastic Programming*, Kluwer, Dordrecht, 1995.

[48] H. Wu, M. Shahidehpour, Z. Li, and W. Tian, "Chance-constrained day-ahead scheduling in stochastic power system operation," *IEEE Transactions on Power Systems*, vol. 29, no. 4, pp. 1583–1591, July 2014.

[49] C. Zhao and Y. Guan, "Unified stochastic and robust unit commitment," *IEEE Transactions on Power Systems*, vol. 28, no. 3, pp. 3353–3361, August 2013.

[50] B. Hu, L. Wu, and M. Marwali, "On the robust solution to SCUC with load and wind uncertainty correlations," *IEEE Transactions on Power Systems*, vol. 29, no. 6, pp. 2952–2964, November 2014.

[51] L. Wu, M. Shahidehpour, and Z. Li, "Comparison of scenario-based and interval optimization approaches to stochastic SCUC," *IEEE Transactions on Power Systems*, vol. 27, no. 2, pp. 913–921, May 2012.

[52] Z. Chen, L. Wu, and M. Shahidehpour, "Effective load carrying capability evaluation of renewable energy via stochastic long-term hourly-based SCUC," *IEEE Transactions on Sustainable Energy*, vol. 6, no. 1, pp. 188–197, January 2015.

[53] B. Hu and L. Wu, "Robust SCUC considering continuous/discrete uncertainties and quick-start units: a two-stage robust optimization with mixed-integer recourse," *IEEE Transactions on Power Systems*, vol. 31, no. 2, pp. 1407–1419, March 2016.

[54] B. Hu and L. Wu, "Robust SCUC with multi-band nodal load uncertainty set," *IEEE Transactions on Power Systems*, vol. 31, no. 3, pp. 2491–2492, 2016.

CHAPTER 6

DAY-AHEAD SCHEDULING: RESERVE DETERMINATION AND VALUATION

Ruiwei Jiang, Antonio J. Conejo, and Jianhui Wang

6.1 THE NEED OF RESERVES FOR POWER SYSTEM OPERATION

In power system operations, reserves are traditionally used to restore the balance between power supply and demand after a power system equipment failure or emergency. Reserves are procured by the system operator (e.g., PJM [1]) as mandated by the regulator (e.g., FERC [2, 3]).

Reserve needs are diverse as they must cover a wide range of unavailabilities and across multiple time-scales. They are generally classified by their response times, which measure how fast the reserved spare capacity of a generating unit can be called upon. Although the categorization of reserves varies by market and system [2, 3], they are more or less similar. However, the increasing number of stochastic production units (wind- or solar-based) entering the production mix calls for a redesign of the reserve products to be procured by the system operator and provided by generating units (see, e.g., References [4–6]).

It is important to note that reserves are normally provided by non-stochastic units that produce energy. In other words, reserve and energy are different commodities provided by the same production facilities: the generators. This implies that decisions on energy production and on reserve allocation must be jointly made to optimally use the production facilities. However, such practice differs in real-world electricity markets; specifically, some European markets generally decide first on energy and then on reserves (e.g., the markets of Spain and Portugal [7]).

As the generation sources of an electric energy system become increasingly stochastic (as a result of adding stochastic production units), the needs for reserves increase in a complex, nonlinear manner. It is therefore important to build and use realistic models to comprehend this complex dependency. These models provide the basis for optimal decision-making on reserve allocation and energy production

Power Grid Operation in a Market Environment: Economic Efficiency and Risk Mitigation, First Edition.
Edited by Hong Chen.
© 2017 by The Institute of Electrical and Electronics Engineers, Inc. Published 2017 by John Wiley & Sons, Inc.

commitments. In the existing literature, deterministic reserve allocation models often impose an arbitrary security margin by enforcing that the total capacity online (or the capacity that can be quickly online) is above $(1 + R\%)$ of the forecasted demand, where $R\%$ is selected based on operator experience. We also refer the readers to a recent survey [8] and the reference therein for reserve allocation models based on stochastic optimization.

Finally, we point out that reserve costs are significant, and for systems with a large percentage of stochastic production, such cost can be an important percentage of the total cost of providing energy in a secure manner, that is, with the appropriate level of reserves.

This chapter provides a tutorial overview on how to optimally procure reserves using both stochastic programming and robust optimization procedures. As the stochastic production units take up a growing share in the electricity production mix, it becomes more important for the Regional Transmission Organizations (RTOs) to better model the stochasticity and accordingly procure reserves. As a result, stochastic programming and robust optimization procedures are particularly well motivated in reserve determination. Additionally, the way in which reserve needs and reserve cost increase with renewable integration and renewable variability is analyzed and illustrated.

6.2   RESERVE DETERMINATION VIA STOCHASTIC PROGRAMMING

In this section, we formulate the reserve determination problem based on two-stage stochastic programming. In the first stage, we schedule the day-ahead commitment of thermal units and corresponding (tentative) generation amounts. Also, we schedule upward and downward reserve capacities for each committed unit to hedge against the uncertain renewable generation. In the second stage, after the renewable uncertain generation is realized, we satisfy the loads by (i) adjusting the actual generation amounts according to the scheduled generation amounts and upward/downward reserve capacities, (ii) curtailing the available renewable generation if needed, and (iii) curtailing the loads if needed. Two-stage stochastic programming has been applied in many areas of energy system operations, including electricity markets (see, e.g., References [9, 10]), reserve requirement (see, e.g., Reference [11]), security-constrained unit commitment (see, e.g., References [12–14]), and wind power utilization (see, e.g., Reference [15]).

6.2.1   Formulation

All notations used in this chapter are summarized in Appendix 6.A. For the reserve determination formulation, we first let $\mathcal{N} := \{1, \ldots, N\}$ represent the set of thermal units in a power grid, and $\mathcal{T} := \{1, \ldots, T\}$ represent the set of operational intervals (e.g., hours). Also, for each generator $n \in \mathcal{N}$ and each time unit $t \in \mathcal{T}$, we let x_{nt} represent the on/off status (i.e., $x_{nt} = 1$ if the generator is on during interval t and $x_{nt} = 0$ otherwise), u_{nt} represent the start-up operation (i.e., $u_{nt} = 1$ if the generator is started up at the beginning of interval t and $u_{nt} - 0$ otherwise), v_{nt} represent the shut-down operation (i.e., $v_{nt} = 1$ if the generator is shut down at the beginning of interval t and $v_{nt} = 0$ otherwise), p_{nt} represent the scheduled generation amount, r_{nt}^{U} represent the scheduled upward reserve amount, r_{nt}^{D} represent the scheduled downward reserve amount, and q_{nt} represent the actual generation amount. Accordingly, we let positive real numbers c_{nt}^{x}, c_{nt}^{u}, c_{nt}^{v}, c_{nt}^{rU}, c_{nt}^{rD}, and c_{nt}^{q} represent the unit cost pertaining to x_{nt}, u_{nt}, v_{nt}, r_{nt}^{U}, r_{nt}^{D}, and q_{nt}, respectively. In this formulation, we assume that the system load D_t is deterministic for each time unit $t \in \mathcal{T}$ and consider the uncertainty pertaining to renewable energy such as wind and solar. To describe the random generation amount from the renewable units, for each time unit $t \in \mathcal{T}$, we let ξ_t represent the total amount of available renewable energy, y_t represent the utilized amount of renewable energy, and w_t represent the curtailed amount of renewable energy. Finally, we let s_t represent the load-shedding amount, and o_t represent the generation curtailment amount in each time unit $t \in \mathcal{T}$. Accordingly, we let c_t^s, c_t^o, and c_t^w represent the unit cost pertaining to s_t, o_t, and w_t, respectively. To avoid trivial cases and without loss of generality, we assume that $c_t^s > c_{nt}^q$ for each $n \in \mathcal{N}$ and $t \in \mathcal{T}$ and $c_t^o > c_t^w$ for each $t \in \mathcal{T}$. Based on this notation, we present the reserve determination model based on stochastic programming as follows.

$$\min_{x,u,v,p,r} \sum_{n=1}^{N} \sum_{t=1}^{T} \left(c_{nt}^{u} u_{nt} + c_{nt}^{v} v_{nt} + c_{nt}^{x} x_{nt} + c_{nt}^{rD} r_{nt}^{D} + c_{nt}^{rU} r_{nt}^{U} \right) + \rho \left[Q(p,r,\xi) \right] \quad (6.1a)$$

$$\text{s.t.} \quad x_{n(t-1)} - x_{nt} + u_{nt} \geq 0, \quad \forall n \in \mathcal{N}, \forall t \in \mathcal{T}, \quad (6.1b)$$

$$v_{nt} = x_{nt} - x_{n(t-1)}, \quad \forall n \in \mathcal{N}, \forall t \in \mathcal{T}, \quad (6.1c)$$

$$x_{nt} - x_{n(t-1)} \leq x_{n\tau}, \quad \forall n \in \mathcal{N}, \forall t \in \mathcal{T}, \forall \tau = t + 1, \dots, \\ \min\{t + MU_n - 1, T\}, \quad (6.1d)$$

$$x_{n(t-1)} - x_{nt} \leq 1 - x_{n\tau}, \quad \forall n \in \mathcal{N}, \forall t \in \mathcal{T}, \forall \tau = t + 1, \dots, \\ \min\{t + MD_n - 1, T\}, \quad (6.1e)$$

$$P_n^{\min} x_{nt} \leq p_{nt} \leq P_n^{\max} x_{nt}, \quad \forall n \in \mathcal{N}, \forall t \in \mathcal{T}, \quad (6.1f)$$

$$p_{nt} + r_{nt}^{U} \leq P_n^{\max} x_{nt}, \quad \forall n \in \mathcal{N}, \forall t \in \mathcal{T}, \quad (6.1g)$$

$$p_{nt} - r_{nt}^{D} \geq P_n^{\min} x_{nt}, \quad \forall n \in \mathcal{N}, \forall t \in \mathcal{T}, \quad (6.1h)$$

$$x, u, v \in \{0, 1\}^{N \times T}, \ p, r^{U}, r^{D} \geq 0, \quad (6.1i)$$

where

$$Q(p, r, \xi) = \min_{q,y,w,s,o} \sum_{n=1}^{N} \sum_{t=1}^{T} \left(c_{nt}^{q} q_{nt} + c_t^s s_t + c_t^o o_t + c_t^w w_t \right) \quad (6.2a)$$

$$\text{s.t.} \quad p_{nt} - r_{nt}^{D} \leq q_{nt} \leq p_{nt} + r_{nt}^{U}, \quad \forall n \in \mathcal{N}, \forall t \in \mathcal{T}, \quad (6.2b)$$

$$y_t + s_t - o_t + \sum_{n=1}^{N} q_{nt} = D_t, \quad \forall t \in \mathcal{T}, \quad (6.2c)$$

$$y_t + w_t = \xi_t, \quad \forall t \in \mathcal{T}, \quad (6.2d)$$

$$q, y, w, s, o \geq 0. \quad (6.2e)$$

In formulations (6.1)–(6.2) described above, we seek to minimize the total cost including scheduling cost and the operating cost $Q(p, r, \xi)$ (corresponding to given variables (p, r) and realized renewable energy ξ) based on a risk measure $\rho(\cdot)$. For example, $\rho(\cdot)$ can be set as the expectation $\mathbb{E}_\xi[\cdot]$ for a risk-neutral power system operator, or as the conditional value-at-risk ($\mathrm{CVaR}[\cdot]$) for a risk-averse one. Constraints (6.1b) (respectively, (6.1c)) describe the generator start-up (respectively, shut-down) operations, constraints (6.1d) (respectively, (6.1e)) describe the generator minimum-up time (respectively, minimum-down time) restrictions, constraints (6.1f) describe bounds of scheduled generation amounts, and constraints (6.1g) (respectively, (6.1h)) describe bounds of upward (respectively, downward) reserve amounts. In addition, constraints (6.2b) describe bounds of actual generation amounts, constraints (6.2c) describe system balance between generation amounts and loads, and constraints (6.2d) describe amounts of renewable utilization and curtailment. Note here that to facilitate analysis and presentation, we simplify the model by relaxing the ramp-rate limits of the thermal generators and transmission line capacity constraints. Note, however, that this relaxation can be removed from the formulation without altering the proposed decision-making model. Based on this relaxation, we only need to consider total amounts of renewable energy, load, and load-shedding (i.e., ξ_t, D_t, and s_t) for each time unit $t \in \mathcal{T}$ in formulation (6.2), because the distributions of these amounts amongst the power grid do not matter.

6.2.2   Formulation Properties

To facilitate understanding and solving formulations (6.1)–(6.2), we analyze below structural properties of the operating cost function $Q(p, r, \xi)$. We first observe that variables p and r, and random variables ξ are fixed parameters in the second-stage problem, and so $Q(p, r, \xi)$ is a (jointly) *convex* function of p, r, and ξ. Furthermore, based on duality theory of linear programming problems [16], we can obtain that $Q(p, r, \xi)$ is convex and *piecewise linear* in variables (p, r, ξ). This property facilitates solving formulations (6.1)–(6.2) based on the Benders' decomposition (see Reference [17]), which in each iteration identifies a linear piece of function $Q(p, r, \xi)$ and accordingly adds it as an additional constraint (denoted as Benders' cut) into the first-stage problem. Next, we represent function $Q(p, r, \xi)$ as a convex and piecewise linear function in an explicit manner. By doing this, we have a clear picture of how the operating cost depends on the scheduled reserve amounts and renewable energy capacity. Also, it helps generate Benders' cuts in a more computationally efficient way.

We characterize the proposed representation as follows.

For each $t \in \mathcal{T}$, let $[1], [2], \ldots, [N]$ represent a permutation of set $\{1, 2, \ldots, N\}$ such that $c_{[1]t}^q \leq c_{[2]t}^q \leq \cdots \leq c_{[N]t}^q$, and define $z_t := D_t - \sum_{n=1}^N (p_{nt} - r_{nt}^D)$ and $r_{nt} := r_{nt}^U + r_{nt}^D$ for each $n \in \mathcal{N}$ and $t \in \mathcal{T}$. Then for any given p, r, and ξ, the operating cost function $Q(p, r, \xi) = \sum_{t=1}^T [Q_t(p, r, \xi) + \sum_{n=1}^N c_{nt}^q (p_{nt} - r_{nt}^D)]$, where

$$Q_t(p,r,\xi) = \begin{cases} -c_t^o z_t + c_t^w \xi_t & \text{if } z_t < 0, \\ -c_t^w(z_t - \xi_t) & \text{if } 0 \le z_t < \xi_t, \\ c_{[k]t}^q(z_t - \xi_t) - \sum_{n=1}^{k-1}\left(c_{[k]t}^q - c_{[n]t}^q\right)r_{[n]t} & \text{if } \xi_t + \sum_{n=1}^{k-1}r_{[n]t} \le z_t < \xi_t + \sum_{n=1}^{k}r_{[n]t}, \\ & \forall k = 1, \dots, N, \\ c_t^s(z_t - \xi_t) - \sum_{n=1}^{N}(c_t^s - c_{nt}^q)r_{nt} & \text{if } z_t \ge \xi_t + \sum_{n=1}^{N}r_{nt}. \end{cases}$$

$$(6.3)$$

This can be shown as follows. For any given variables (p, r, ξ), we obtain $Q(p, r, \xi)$ by solving the second-stage problem (6.2). First, in view that (i) formulation (6.2) can be decomposed into T sub-problems for each time unit $t \in \mathcal{T}$, and (ii) $q_{nt} \ge p_{nt} - r_{nt}^D$ due to constraints (6.2b), we can reformulate the second-stage problem (6.2) as

$$Q(p,r,\xi) = \sum_{t=1}^{T}\sum_{n=1}^{N}c_{nt}^q(p_{nt} - r_{nt}^D) + \sum_{t=1}^{T}\left\{ \min_{q',y,w,s,o\ge 0} \sum_{n=1}^{N}\left(c_{nt}^q q_{nt}' + c_t^s s_t + c_t^o o_t + c_t^w w_t\right)\right\}$$

$$(6.4a)$$

$$\text{s.t. } 0 \le q_{nt}' \le r_{nt}, \quad \forall n \in \mathcal{N}, \quad (6.4b)$$

$$y_t + s_t - o_t + \sum_{n=1}^{N}q_{nt}' = z_t, \quad (6.4c)$$

$$y_t + w_t = \xi_t, \quad (6.4d)$$

where we denote $q_{nt}' := q_{nt} - (p_{nt} - r_{nt}^D)$ and $Q_t(p, r, \xi)$ as the optimal objective value of the embedded optimization problem in formulation (6.4) for each $t \in \mathcal{T}$. In the following, we use superscript $*$ to denote optimal solutions (e.g., o_t^*), and solve the embedded optimization problem by discussing the following four cases:

Case 1. If $z_t < 0$, then there are excess generation scheduled and insufficient downward reserve. It follows that $o_t^* = -z_t$, $y_t^* = s_t^* = q_{nt}'^* = 0$, and $w_t^* = \xi_t$ for each $n \in \mathcal{N}$ due to constraint (6.4c), $c_t^o > c_t^w$ and $q', y, w, s, o \ge 0$. It follows that $Q_t(p, r, \xi) = -c_t^o z_t + c_t^w \xi_t$.

Case 2. If $0 \le z_t < \xi_t$, then the system load can be covered by the minimum scheduled generation plus the renewable energy. It follows that $o_t^* = 0$ due to constraint (6.4c), $z_t \ge 0$ and $c_t^o > c_t^w$. It follows that $y_t + s_t + \sum_{n=1}^{N}q_{nt}' = z_t$. Since $c_{nt}^q, c_t^s > 0$ and the unit cost of y_t is zero, we have $y_t^* = z_t$, $w_t^* = \xi_t - z_t$ and $s_t^* = q_{nt}'^* = 0$ for each $n \in \mathcal{N}$. Note here that y_t^* and w_t^* are feasible because $z_t < \xi_t$ and constraint (6.4d) is satisfied. Hence, $Q_t(p, r, \xi) = -c_t^w(z_t - \xi_t)$.

Case 3. For each $k = 1, \ldots, N$, if $\xi_t + \sum_{n=1}^{k-1} r_{[n]t} \le z_t < \xi_t + \sum_{n=1}^{k} r_{[n]t}$, then the system load can be covered by the minimum scheduled generation plus the renewable energy, together with upward reserves from the k cheapest generators. It follows that $o_t^* = 0$ due to constraint (6.4c), $z_t \ge 0$ and $c_t^o > 0$.

It follows that $y_t + s_t + \sum_{n=1}^{N} q'_{nt} = z_t$. In view that the unit costs of y_t, q'_{nt} and s_t can be ordered as $0 < c_{[1]t}^q \le c_{[2]t}^q \le \cdots \le c_{[N]t}^q < c_t^s$ by assumption, we can optimally break down the value of z_t into y_t, q'_{nt} and s_t by

$$
y_t^* = \xi_t, \quad q'^*_{[n]t} = \begin{cases} r_{[n]t}, & \text{if } n = 1, \ldots, k-1, \\ z_t - \left(\xi_t + \sum_{n=1}^{k-1} r[n]t \right), & \text{if } n = k, \\ 0, & \text{if } n = k, \ldots, N, \end{cases} \quad s_t^* = 0.
$$

Intuitively, this assignment of z_t fills in y_t, q'_{nt}, and s_t one after another in a nondecreasing order of their unit costs, while respecting their upper and lower bounds, that is, $0 \le y_t \le \xi_t, 0 \le q'_{[n]t} \le r_{[n]t}$ for each $n \in \mathcal{N}$, and $s_t \ge 0$. It follows that

$$
\begin{aligned}
Q_t(p, r, \xi) &= \sum_{n=1}^{N} \left(c_{nt}^q q'^*_{nt} + c_t^s s_t^* + c_t^o o_t^* \right) \\
&= \sum_{n=1}^{k-1} c_{[n]t}^q r_{[n]t} + c_{[k]t}^q \left(z_t - \xi_t - \sum_{n=1}^{k-1} r_{[n]t} \right) \\
&= c_{[k]t}^q (z_t - \xi_t) - \sum_{n=1}^{k-1} (c_{[k]t}^q - c_{[n]t}^q) r_{[n]t}.
\end{aligned}
$$

Case 4. If $z_t \ge \xi_t + \sum_{n=1}^{N} r_{nt}$, then there are unsatisfied system load even if all upward reserves are utilized. It follows that $o_t^* = 0$ due to constraint (6.4c), $z_t \ge 0$ and $c_t^o > 0$. It follows that $y_t + s_t + \sum_{n=1}^{N} q'_{nt} = z_t$. In view that $y_t + \sum_{n=1}^{N} q'_{nt} \le \xi_t + \sum_{n=1}^{N} r_{nt} \le z_t$ and $c_t^s > c_{nt}^q$ for each $n \in \mathcal{N}$, we can optimally break down the value of z_t into y_t, q'_{nt}, and s_t by

$$
y_t^* = \xi_t, \quad q'^*_{nt} = r_{nt}, \quad \forall n \in \mathcal{N}, \quad s_t^* = z_t - \xi_t - \sum_{n=1}^{N} r_{nt}.
$$

It follows that

$$
\begin{aligned}
Q_t(p, r, \xi) &= \sum_{n=1}^{N} \left(c_{nt}^q q'^*_{nt} + c_t^s s_t^* + c_t^o o_t^* \right) \\
&= \sum_{n=1}^{N} c_{nt}^q r_{nt} + c_t^s \left(z_t - \xi_t - \sum_{n=1}^{N} r_{nt} \right) \\
&= c_t^s (z_t - \xi_t) - \sum_{n=1}^{N} (c_t^s - c_{nt}^q) r_{nt}.
\end{aligned}
$$

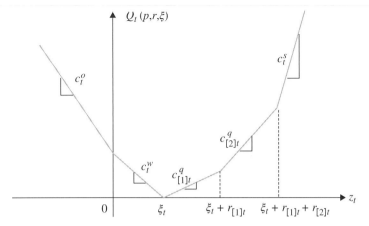

Figure 6.1   An example of function $Q_t(p, r, \xi)$ with $N = 2$.

As an example, we depict function $Q_t(p, r, \xi)$ as a function of z_t in Figure 6.1 with $N = 2$. In this figure, it can be observed that $Q_t(p, r, \xi)$ is convex and piecewise linear as presented in equation (6.3). Also, in view that $Q_t(p, r, \xi)$ is continuous, convex and piecewise linear, we can represent $Q_t(p, r, \xi)$ as

$$Q_t(p, r, \xi) = \max \left\{ \begin{array}{l} -c_t^o z_t + c_t^w \xi_t, \\[2mm] -c_t^w(z_t - \xi_t), \\[2mm] c_{[k]t}^q(z_t - \xi_t) - \sum_{n=1}^{k-1}(c_{[k]t}^q - c_{[n]t}^q)r_{[n]t}, \quad \forall k = 1, \ldots, N, \\[2mm] c_t^s(z_t - \xi_t) - \sum_{n=1}^{N}(c_t^s - c_{nt}^q)r_{nt} \end{array} \right\}.$$

6.2.3 Solution Algorithm

In practical applications, we usually solve the risk-neutral stochastic model (6.1) (i.e., when $\rho(\cdot) = \mathbb{E}[\cdot]$) by using the sample average approximation (SAA) method, which employs the sample average value $(1/M)\sum_{m=1}^{M} Q(p, r, \xi^m)$ as a surrogate of the expectation $\mathbb{E}[Q(p, r, \xi)]$, where $\{\xi^m : m = 1, \ldots, M\}$ represents a sequence independent and identically distributed (i.i.d.) samples of ξ. Accordingly, the SAA surrogate of the stochastic model (6.1) is

$$\min_{x,u,v,p,r,Q} \sum_{n=1}^{N}\sum_{t=1}^{T}\left(c_{nt}^u u_{nt} + c_{nt}^v v_{nt} + c_{nt}^x x_{nt} + c_{nt}^{rD} r_{nt}^D + c_{nt}^{rU} r_{nt}^U + c_{nt}^q(p_{nt} - r_{nt}^D)\right)$$

$$+\frac{1}{M}\sum_{m=1}^{M}\sum_{t=1}^{T} Q_t^m \quad \text{s.t. } (6.1b)-(6.1i), \tag{6.5}$$

where Q_t^m is a decision variable representing the value of $Q_t(p, r, \xi^m)$. By exploiting the structural properties of function $Q_t(p, r, \xi^m)$, we can employ the Benders' decomposition algorithm (see Reference [17]) to solve the SAA surrogate. We summarize the algorithm as follows.

0. Input: lower bound LB := $-\infty$, upper bound UB := $+\infty$, optimality gap tolerance ϵ, iteration number limit L, set of Benders' cuts CUT := $\emptyset$.

1. For $\ell = 1, \dots, L$, repeat the following steps:

    (1) Solve the SAA surrogate (6.5) with the current set of Benders' cuts in CUT as additional constraints. Record optimal solutions (p^*, r^*, Q^*), and set LB equal to the optimal objective value.

    (2) For $m = 1, \dots, M$ and $t = 1, \dots, T$, repeat the following steps:

        (a) Sort $\{1, \dots, N\}$ to generate a permutation $[1], \dots, [N]$ as defined in Section 6.2.2.

        (b) Compute the value of $Q_t(p^*, r^*, \xi^m)$ using (6.3).

        (c) Check if $Q_t^{m*} \geq Q_t(p^*, r^*, \xi^m)$. If yes, continue; if no, identify a violated piece of $Q_t(p^*, r^*, \xi^m)$ and add a corresponding Benders' cut in the form of $Q_t^m \geq Q_t(p^*, r^*, \xi^m)$ into set CUT. For example, if $Q_t^{m*} < Q_t(p^*, r^*, \xi^m)$ and $\xi_t^m + \sum_{n=1}^{k-1}(r_{nt}^{U*} + r_{nt}^{D*}) \leq D_t - \sum_{n=1}^{N}(p_{nt}^* - r_{nt}^{D*}) < \xi_t^m + \sum_{n=1}^{k}(r_{nt}^{U*} + r_{nt}^{D*})$ for some $k = 1, \dots, N$, add a cut

$$Q_t^m \geq c_{[k]t}^q \left(D_t - \sum_{n=1}^{N}(p_{nt} - r_{nt}^D) - \xi_t^m \right) - \sum_{n=1}^{k-1}(c_{[k]t}^q - c_{[n]t}^q)\left(r_{[n]t}^U + r_{[n]t}^D \right).$$

    (3) Set UB equal to LB + $\sum_{m=1}^{M}\sum_{t=1}^{T}\left(Q_t(p^*, r^*, \xi^m) - Q_t^{m*}\right)$.

    (4) If $|UB - LB|/LB < \epsilon$ or no cuts are added to CUT, then return and output (p^*, r^*) as an optimal solution; otherwise, go to Step 1.(1).

6.2.4 Illustrative Example

We illustrate the proposed procedure using a simple example based on a two-bus power system (see Figure 6.2) containing one conventional generator C and one renewable generator R. For simplicity, we omit subscripts n and t, and assume one time interval, unit costs $c^q = \$2/MWh$, $c^{rU} = c^{rD} = \$1/MW$, penalty costs $c^s = c^o = 2c^w = \$10/MW$, and generation bounds $P^{max} = 10P^{min} = 10MW$. Also,

Figure 6.2   A two-bus power system example.

we assume that $D = 9$MW and the available amount of renewable energy ξ has a distribution function $F(\cdot)$. Considering (6.3), we obtain that function $Q(p, r, \xi)$ can be represented as

$$Q(p, r, \xi) = c^q(p - r^D) + \begin{cases} -c^o z + c^w \xi & \text{if } z < 0, \\ -c^w(x - \xi) & \text{if } 0 \leq z < \xi, \\ c^q(z - \xi) & \text{if } \xi \leq z < \xi + (r^D + r^U), \\ c^s(z - \xi) - (c^s - c^q)(r^D + r^U) & \text{if } z \geq \xi + (r^D + r^U), \end{cases}$$

(6.6)

where $z = D - p + r^D$. From the representation (6.6), it is clear that all solutions with $z < 0$ are dominated by solutions with $z = 0$ because $-c^o z + c^w \xi > -c^w(0 - \xi)$ when $z < 0$. Hence, we can assume that $z \geq 0$ without loss of generality. It follows that the risk-neutral reserve determination model (6.1) can be reformulated as

$$\min_{p, r^D, r^U} G(p, r^D, r^U) := c^{rD} r^D + c^{rU} r^U + \mathbb{E}[Q(p, r, \xi)]$$

(6.7)

$$\text{s.t. (6.1b)} - \text{(6.1i)},$$

where we let $G(p, r^D, r^U)$ represent the objective function of formulation (6.7) and

$$\mathbb{E}[Q(p, r, \xi)] = c^q(p - r^D) + \int_0^{z - r^D - r^U} \left[c^s(z - \xi) - (c^s - c^q)(r^D + r^U) \right] dF(\xi) +$$

$$\int_{z - r^D - r^U}^z c^q(z - \xi) dF(\xi) - \int_z^\infty c^w(z - \xi) dF(\xi)$$

based on (6.6). Note here that there is only one conventional generator in the two-bus system, and hence the generator is on (i.e., $x = 1$) and constraints (6.1b)–(6.1e) are satisfied. Also note that the remaining constraints 6.1f–6.1h can be combined as

$$P^{\min} \leq p - r^D, \quad p + r^U \leq P^{\max}.$$

(6.8)

To solve model (6.7), we relax constraint (6.8) and solve an unconstrained version of model (6.7). We will show how to transform the obtained optimal solution to satisfy (6.8) later on. To solve the unconstrained version of model (6.7), we take the first derivatives of $G(p, r^D, r^U)$ to obtain

$$\frac{\partial G}{\partial p}(p, r^D, r^U) = c^q + c^w - (c^q + c^w)F(D - p + r^D) - (c^s - c^q)F(D - p - r^U),$$

$$\frac{\partial G}{\partial r^D}(p, r^D, r^U) = (c^{rD} - c^q - c^w) + (c^q + c^w)F(D - p + r^D),$$

$$\frac{\partial G}{\partial r^U}(p, r^D, r^U) = c^{rU} - (c^s - c^q)F(D - p - r^U).$$

Since $G(p, r^D, r^U)$ is jointly convex in variables (p, r^D, r^U), we can let $\frac{\partial G}{\partial r^U}(p, r^D, r^U) = \frac{\partial G}{\partial r^D}(p, r^D, r^U) = 0$ to yield

$$p - r^D = D - F^{-1}\left(\frac{c^w + c^q - c^{rD}}{c^w + c^q}\right), \quad p + r^U = D - F^{-1}\left(\frac{c^{rU}}{c^s - c^q}\right),$$

which further leads to

$$\frac{\partial G}{\partial p}(p, r^D, r^U) = c^{rD} - c^{rU} = 0.$$

Therefore, at optimality, we have

$$p^* - r^{D*} = \max\left\{P^{\min}, D - F^{-1}\left(\frac{c^w + c^q - c^{rD}}{c^w + c^q}\right)\right\} = \max\left\{1, 9 - F^{-1}\left(\frac{6}{7}\right)\right\},$$

$$p^* + r^{U*} = \min\left\{P^{\max}, D - F^{-1}\left(\frac{c^{rU}}{c^s - c^q}\right)\right\} = \min\left\{10, 9 - F^{-1}\left(\frac{1}{8}\right)\right\},$$

and p^* can be any values within interval $[p^* - r^{D*}, p^* + r^{U*}]$ because $\partial G / \partial p(p, r^D, r^U) = 0$. Note that the $\max\{\cdot\}$ and $\min\{\cdot\}$ operations guarantee that constraint (6.8) is satisfied at optimality. For example, suppose that $F(\cdot)$ has a support $[2\text{MW}, 4\text{MW}]$, that is, $F(2) = 0$ and $F(4) = 1$. In Table 6.1, we display the optimal solutions $p^* - r^{D*}$ and $p^* + r^{U*}$ with $F(\cdot)$ having various distributions and parameter settings. In each setting, we also compare the optimal solutions with the case without reserve (i.e., r^D and r^U are fixed at zeros). From Table 6.1, we can observe that the optimal generation amount p^* without reserve is always within the interval $[p^* - r^{D*}, p^* + r^{U*}]$ when reserves are available. Note that, in Table 6.1, we display the mean and variance of the Weibull random variables for a clear comparison with the corresponding Normal random variables. The scale and

TABLE 6.1 Optimal reserve determinations under various distributions and parameter settings

Distribution	Parameters	p^* without Reserve (MW)	$p^* - r^{D*}$ (MW)	$p^* + r^{U*}$ (MW)
Uniform	Range [2, 4]	6.07	5.29	6.75
Normal	mean $= 3$, s.d. $= 1/3$	6.03	5.65	6.38
	mean $= 3$, s.d. $= 1/2$	6.04	5.50	6.54
	mean $= 3$, s.d. $= 1$	6.06	5.34	6.69
Weibull	mean $= 3$, s.d. $= 1/3$	6.03	5.64	6.40
	mean $= 3$, s.d. $= 1/2$	6.12	5.53	6.57
	mean $= 3$, s.d. $= 1$	6.49	5.65	6.89
Beta	$\alpha = 0.1, \beta = 0.5$	7.00	5.92	7.00
	$\alpha = 0.3, \beta = 0.5$	6.59	5.20	6.99
	$\alpha = 0.5, \beta = 0.5$	6.10	5.10	6.92
	$\alpha = 0.5, \beta = 0.3$	5.61	5.01	6.84
	$\alpha = 0.5, \beta = 0.1$	5.01	5.00	6.24

shape parameters of the Weibull random variables are (scale $= 3.14$, shape $= 10.88$), (scale $= 3.21$, shape $= 7.06$), and (scale $= 3.34$, shape $= 3.29$) respectively.

6.3 RESERVE DETERMINATION VIA ADAPTIVE ROBUST OPTIMIZATION

Complementarily, in this section, we formulate the reserve determination problem based on two-stage adaptive robust optimization. As compared to the stochastic reserve determination model (6.1), the robust reserve determination model attempts to protect against the worst-case renewable generation scenario (within an uncertainty set) where the highest operational cost is incurred, and minimize the total costs when the worst-case scenario is realized. Recently, robust optimization models have received increasing attentions in the energy system literature. For example, they have been applied in building offering curves (see, e.g., Reference [18]) and unit commitment with renewable energy integration (see, e.g., References [19–21]), among others.

6.3.1 Formulation

The robust reserve determination model is formulated as follows.

$$\min_{x,u,v,p,r} \sum_{n=1}^{N} \sum_{t=1}^{T} \left(c_{nt}^{u} u_{nt} + c_{nt}^{v} v_{nt} + c_{nt}^{x} x_{nt} + c_{nt}^{rD} r_{nt}^{D} + c_{nt}^{rU} r_{nt}^{U} \right) + \max_{\xi \in \Omega} \; Q(p,r,\xi)$$

$$\text{s.t. } (6.1b) - (6.1i), \tag{6.9}$$

where $Q(p,r,\xi)$ is defined in formulation (6.2). On the one hand, as compared to the stochastic reserve determination model (6.1), the robust model (6.9) captures the uncertainty of ξ by using a physically meaningful uncertainty set Ω. On the other hand, the robust model is distribution-free, that is, Ω can be developed without assuming or accurately estimating the probability distribution of ξ. For example, with a minimal amount of data, Ω can be developed as the following polyhedral uncertainty set

$$\Omega = \left\{ \xi \in \mathbb{R}^{T} : \xi_{t}^{L} \leq \xi_{t} \leq \xi_{t}^{U}, \sum_{t=1}^{T} \xi_{t} \geq \overline{\xi}_{0} \right\},$$

where ξ_{t}^{L} and ξ_{t}^{U} represent lower and upper bounds of available renewable energy amount in time unit t respectively, and $\overline{\xi}_{0}$ represents a lower bound of total available renewable energy amount in the operational time intervals. The values of ξ_{t}^{L}, ξ_{t}^{U}, and $\overline{\xi}_{0}$ can be conveniently estimated from statistical inferences of ξ. For example, we can set ξ_{t}^{L} and ξ_{t}^{U} as the 10th and 90th percentiles of ξ_{t} respectively, and $\overline{\xi}_{0}$ as the 10th percentile of $\sum_{t=1}^{T} \xi_{t}$.

To solve the robust model (6.9), we reformulate the second-stage problem as a bilinear program based on the strong duality of linear program [16]. More specifically, we can dualize the linear program (6.2) to yield the following reformulation:

$$
\max_{\xi \in \Omega} Q(p, r, \xi) = \max_{\xi, \lambda^{\pm}, \mu, \theta} \sum_{t=1}^{T} \left[\sum_{n=1}^{N} (p_{nt} - r_{nt}^{D}) \lambda_{nt}^{+} \right.
$$

$$
\left. - \sum_{n=1}^{N} (p_{nt} + r_{nt}^{U}) \lambda_{nt}^{-} + D_t \mu_t + \xi_t \theta_t \right] \tag{6.10a}
$$

$$
\text{s.t.} \quad \lambda_{nt}^{+} - \lambda_{nt}^{-} + \mu_t \geq c_{nt}^{q}, \quad \forall n \in \mathcal{N}, \ \forall t \in \mathcal{T}, \tag{6.10b}
$$

$$
\mu_t + \theta_t \leq 0, \quad \forall t \in \mathcal{T}, \tag{6.10c}
$$

$$
\theta_t \leq c_{nt}^{w}, \quad \forall t \in \mathcal{T}, \tag{6.10d}
$$

$$
-c_t^{o} \leq \mu_t \leq c_t^{s}, \quad \forall t \in \mathcal{T}, \tag{6.10e}
$$

$$
\xi_t^{L} \leq \xi_t \leq \xi_t^{U}, \quad \forall t \in \mathcal{T}, \tag{6.10f}
$$

$$
\sum_{t=1}^{T} \xi_t \geq \overline{\xi}_0, \tag{6.10g}
$$

$$
\lambda_{nt}^{\pm} \geq 0, \quad \forall n \in \mathcal{N}, \ \forall t \in \mathcal{T}, \tag{6.10h}
$$

where $\lambda_{nt}^{\pm}$, μ_t and θ_t represent the dual variables of constraints (6.2b), (6.2c), and (6.2d) respectively.

6.3.2   Solution Algorithm

The reformulated second-stage problem (6.10) is a bilinear program because of the bilinear products $\xi_t \theta_t$ in the objective function (6.10a). In general, a bilinear program in the form of problem (6.10) is NP-hard. In practice, several reformulations and solution algorithms have been applied to solve problem (6.10). On one hand, we can reformulate problem (6.10) to be a mixed-integer linear program (see, e.g., References [19, 21–23]). The reformulation method introduces additional binary variables to represent variable ξ, and linearize problem (6.10) based on the McCormick relaxation [24]. On the other hand, bilinear heuristic (see, e.g., References [20, 21]) has been very helpful in practical applications to find a local optima of problem (6.10) in a short time, which can then be used to generate a Benders' cut. We summarize a simple bilinear heuristic for solving problem (6.10) as follows.

0. Input: a starting point $\xi^0 \in \Omega$, optimality gap tolerance ϵ. Set iteration counter $j = 1$.

1. Fix variable ξ to be ξ^{j-1} and solve problem (6.10) as a linear program. Record optimal solution θ^j and optimal objective value Q_1^j.

2. Fix variable θ to be θ^j and solve problem (6.10) as a linear program. Record optimal solution ξ^j and optimal objective value Q_2^j.

3. If $|Q_2^j - Q_1^j|/Q_2^j < \epsilon$, then a local optima is researched. Fix variable ξ to be ξ^j and solve problem (6.10) as a linear program, record optimal solution $(\lambda^{\pm*}, \mu^*, \theta^*)$, return and output $(\xi^j, \lambda^{\pm*}, \mu^*, \theta^*)$ as a locally optimal solution to problem (6.10). Otherwise, update j to be $j + 1$ and go to Step 1.

Based on solution algorithms of problem (6.10), we can solve the robust reserve determination model (6.9) by the Benders' decomposition algorithm summarized as follows.

0. Input: lower bound LB $:= -\infty$, upper bound UB $:= +\infty$, optimality gap tolerance ϵ, iteration number limit L, set of Benders' cuts CUT $:= \emptyset$.

1. For $\ell = 1, \ldots, L$, repeat the following steps:

(1) Solve the master problem

$$
\min_{x,u,v,p,r,Q} \sum_{n=1}^{N} \sum_{t=1}^{T} \left(c_{nt}^u u_{nt} + c_{nt}^v v_{nt} + c_{nt}^x x_{nt} + c_{nt}^{rD} r_{nt}^D + c_{nt}^{rU} r_{nt}^U \right) + Q
$$

s.t. (6.1b) $-$ (6.1i)

with Q as a decision variable representing the value of $\max_{\xi \in \Omega} Q(p, r, \xi)$, and the current set of Benders' cuts in CUT as additional constraints. Record optimal solutions (p^*, r^*, Q^*), and set LB equal to the optimal objective value.

(2) Solve the reformulated second-stage problem (6.10). Record optimal solutions $(\xi^*, \lambda^{\pm*}, \mu^*, \theta^*)$ and optimal objective value V_{SSP}, and set UB equal to LB $- Q^* + V_{SSP}$.

(3) If $|UB - LB|/LB < \epsilon$ or $Q^* \geq V_{SSP}$, then return and output (p^*, r^*) as an optimal solution; otherwise, go to the next step.

(4) Add a Benders' cut

$$
Q \geq \sum_{t=1}^{T} \left[\sum_{n=1}^{N} (p_{nt} - r_{nt}^D) \lambda_{nt}^{+*} - \sum_{n=1}^{N} (p_{nt} + r_{nt}^U) \lambda_{nt}^{-*} + D_t \mu_t^* + \xi_t^* \theta_t^* \right]
$$

into set CUT.

6.3.3 Illustrative Example

We continue discussing the illustrative example described in Section 6.2.4 in a robust setting, that is, we model the uncertainty of ξ based only on its range $[\xi^L, \xi^U]$ and make no other assumptions on its probability distribution. From the reformulation (6.10), we can observe that the objective function (6.10a) is convex in variable ξ, and so at least one optimal solution ξ^* belong to $\{\xi^L, \xi^U\}$. In other words, we only need to consider ξ^L and ξ^U (rather than the entire interval $[\xi^L, \xi^U]$) for worst-case scenarios.

TABLE 6.2 Optimal robust reserve determinations under various parameter settings

ξ^L (MW)	ξ^U (MW)	p^* without Reserve (MW)	$p^* - r^{D*}$ (MW)	$p^* + r^{U*}$ (MW)
1.5	4.0	6.67	5.71	7.50
2.0	3.5	6.50	5.93	7.00
2.0	4.0	6.33	5.57	7.00
2.0	4.5	6.17	5.21	7.00
2.5	4.0	6.00	5.43	6.50

Based on this observation and equation (6.6), we can formulate the robust model in this example as

$$\min_{p,r^D,r^U} c^{rD}r^D + c^{rU}r^U + Q$$

$$\text{s.t. (6.1f)} - (6.1i), \tag{6.11a}$$

$$Q \geq Q^i, \quad \forall i = L, U, \tag{6.11b}$$

$$Q^i \geq c^q(p - r^D) - c^w(D - p + r^D - \xi^i), \quad \forall i = L, U, \tag{6.11c}$$

$$Q^i \geq c^q(p - r^D) + c^q(D - p + r^D - \xi^i), \quad \forall i = L, U, \tag{6.11d}$$

$$Q^i \geq c^q(p - r^D) + c^s(D - p + r^D - \xi^i) - (c^s - c^q)(r^D + r^U), \quad \forall i = L, U, \tag{6.11e}$$

where Q^L and Q^U represent the values of $Q(p, r, \xi^L)$ and $Q(p, r, \xi^U)$ respectively, and constraints (6.11c)–(6.11e) follow from equation (6.6).

In Table 6.2, we display the optimal solutions $p^* - r^{D*}$ and $p^* + r^{U*}$ with various ranges $[\xi^L, \xi^U]$. Similar to the example for the stochastic model discussed in Section 6.2, any value within the interval $[p^* - r^{D*}, p^* + r^{U*}]$ is optimal for p^*. In each setting, we also compare the optimal solutions with the case without reserve (i.e., r^D and r^U are fixed at zeros). From Table 6.2, similar to the numerical results we obtained in Table 6.1, we can observe that the optimal generation amount p^* without reserve is always within the interval $[p^* - r^{D*}, p^* + r^{U*}]$ when reserves are available.

6.4 STOCHASTIC PROGRAMMING VS. ADAPTIVE ROBUST OPTIMIZATION

In this section, we compare the stochastic and robust reserve determination models described in Sections 6.2 and 6.3, respectively. Clearly, the stochastic reserve determination model (6.1) is less conservative than the robust one (6.9), because for any given generation amounts p, reserve amounts r, and risk measure $\rho(\cdot)$ (e.g., $\rho(\cdot) = \mathbb{E}[\cdot]$), we have

$$\rho(Q(p, r, \xi)) \leq \max_{\xi \in \Omega} Q(p, r, \xi).$$

This observation is also intuitive from the perspective of distributional information utilized in both models. The stochastic model usually utilizes a complete set of distributional information about ξ. For example, before evaluating the values of $\mathbb{E}[Q(p, r, \xi)]$ or $\mathrm{CVaR}(Q(p, r, \xi))$, we need to (accurately) estimate the probability distribution of ξ based on its distribution function or density function. In contrast, all distributional information the robust model utilizes is the range of ξ, and the robust model takes all possible probability distributions of ξ into account as long as they reside on the given range. In other words, as compared to the stochastic model, the robust model utilizes a minimal amount of distributional information which could lead to significant conservatism in reserve determination. Therefore, the stochastic model is more suitable than the robust one when sufficient distributional information is available, for example, when we have a sufficient amount of data to accurately estimate the probability distribution of ξ.

One the other hand, the robust model could be preferable to the stochastic one when the distributional information is limited, which can happen more than often in practice. For example, we may have very few data of ξ when the wind farm or solar system is newly built, or when the weather pattern is varying so fast that only very recent data of are relevant/reliable. In many cases, the accurate estimation of the probability distribution of ξ is unavailable and only descriptive statistics (e.g., range, mean, or variance) can be relied on when determining reserve amounts. Under such circumstances, the stochastic model becomes questionable because (i) we do not know how to estimate the underlying probability distribution based on a limited amount of distributional information, and (ii) it may be challenging to generate a large number of samples for the SAA method. In contrast, the robust model provides a conservative but reliable guarantee for the reserve determination because it demands a minimal amount of distributional information, and meanwhile protects against the worst-case scenarios.

6.4.1 Illustrative Example

We compare the stochastic and robust optimal reserve determination solutions in various parameter settings based on the example we discussed in Sections 6.2.4 and 6.3.3. For the robust approach, we set $\xi^L = 2.0$MW and $\xi^U = 4.0$MW. First, we compare the stochastic and robust optimal solutions in various distributional settings based on an out-of-sample simulation and summarize the results in Table 6.3. More specifically, in each distributional setting, we fix the stochastic and robust solutions obtained in Tables 6.1 and 6.2, and evaluate their average costs based on 1000 i.i.d. samples generated from the corresponding distributional setting. From Table 6.3, we can observe that the average cost of the robust optimal solution is higher than that of the stochastic one in all distributional settings. Also, we can observe that the relative increases of the average cost (see the last column of Table 6.3) are reasonable in most settings, except when the probability distribution of ξ is heavily right-skewed (see, e.g., Beta$(0.5, 0.3)$ and Beta$(0.5, 0.1)$).

Second, we compare the stochastic and robust optimal solutions in worst-case scenarios based on an out-of-sample simulation and summarize the results in

TABLE 6.3 **Comparison of stochastic and robust optimal solutions in various distributional settings**

Distribution	Parameters	Stochastic	Robust	Performance Difference (%)
Uniform	Range $[2, 4]$	13.73	13.98	1.82
Normal	mean $= 3$, s.d. $= 1/3$	13.07	13.54	3.60
	mean $= 3$, s.d. $= 1/2$	13.42	13.71	2.16
	mean $= 3$, s.d. $= 1$	13.67	13.89	1.61
Weibull	mean $= 3$, s.d. $= 1/3$	13.04	13.53	3.76
	mean $= 3$, s.d. $= 1/2$	13.52	13.80	2.07
	mean $= 3$, s.d. $= 1$	14.19	14.26	0.49
Beta	$\alpha = 0.1, \beta = 0.5$	14.98	15.03	0.33
	$\alpha = 0.3, \beta = 0.5$	14.39	14.59	1.39
	$\alpha = 0.5, \beta = 0.5$	14.00	14.39	2.79
	$\alpha = 0.5, \beta = 0.3$	13.52	14.50	7.25
	$\alpha = 0.5, \beta = 0.1$	12.35	14.92	20.81

Table 6.4. More specifically, for each stochastic optimal solution obtained under the corresponding distributional setting, we fix the solution and evaluate its cost under worst-case scenarios. From Table 6.4, we can observe that the worst-case costs of the robust optimal solutions are the same in all distributional settings, verifying that the robust model does not utilize the corresponding distributional information. Also, we can observe that the worst-case cost of the stochastic optimal solution is higher than that of the robust one in all distributional settings. In addition, we can observe that the relative increase of the worst-case cost is nontrivial in most settings, indicating that stochastic optimal solutions can perform poorly when distributional information is limited.

TABLE 6.4 **Comparison of stochastic and robust optimal solutions in worst-case scenarios**

Distribution	Parameters	Stochastic	Robust	Performance Difference (%)
Uniform	Range $[2, 4]$	17.46	15.43	13.16
Normal	mean $= 3$, s.d. $= 1/3$	19.68	15.43	27.54
	mean $= 3$, s.d. $= 1/2$	18.75	15.43	21.52
	mean $= 3$, s.d. $= 1$	17.80	15.43	15.36
Weibull	mean $= 3$, s.d. $= 1/3$	19.60	15.43	27.03
	mean $= 3$, s.d. $= 1/2$	18.46	15.43	19.64
	mean $= 3$, s.d. $= 1$	16.12	15.43	4.47
Beta	$\alpha = 0.1, \beta = 0.5$	17.52	15.43	13.55
	$\alpha = 0.3, \beta = 0.5$	15.84	15.43	2.66
	$\alpha = 0.5, \beta = 0.5$	16.43	15.43	6.48
	$\alpha = 0.5, \beta = 0.3$	17.08	15.43	10.69
	$\alpha = 0.5, \beta = 0.1$	21.32	15.43	38.17

6.5 RESERVE VALUATION

In this section, we conduct numerical simulations of both stochastic and robust reserve determination models on a modified IEEE 30-bus system (see http://www.ee. washington.edu/research/pstca/pf30/pg_tca30bus.htm for the original test case) to evaluate the expected cost saving a power system can benefit from when stochastic units are integrated and reserves are optimally decided. More specifically, we inves-tigate the reserve valuation by comparing the stochastic and robust reserve determi-nation models with their counterparts where reserves are heuristically fixed, respec-tively. The reserve valuation is conducted under various parameter settings that can be summarized in two categories: (1) as a function of the integration level of stochas-tic units, and (2) as a function of the variability of stochastic units. We display the simulation results for the first category in Section 6.5.1, and the simulation results for the second category in Section 6.5.2.

Before displaying the simulation results, we describe the simulation testbed based on the modified IEEE 30-bus system. First, we test for a 24-period oper-ational interval, that is, $T = 24$. Second, the modified IEEE 30-bus system con-tains 10 conventional units whose characteristics are summarized in Table 6.5, and a solar unit. In the simulations, we assume that the renewable energy ξ_t from the solar unit in each time interval t follows a Normal distribution with mean $\mu_t :=$ $\gamma_t \text{INT\%}[(1/T) \sum_{t=1}^{T} D_t]$ and standard deviation $\sigma_t := \text{VAR\%}\mu_t$, where INT% rep-resents the integration level of the solar unit (e.g., INT% = 20%) as compared to the average system load $(1/T) \sum_{t=1}^{T} D_t$, VAR% represents the variability of the solar unit (e.g., VAR% = 25%), and γ_t represents relative solar energy in each time interval t as compared to its peak value and is summarized in Table 6.6. Third, we assume that the system load in each time interval is deterministic as displayed in Table 6.7. Finally, we assume that $c_t^o = \$25/\text{MW}$, $c_t^w = \$20/\text{MW}$ and $c_t^s = \$200/\text{MW}$ for all time intervals $t \in \mathcal{T}$.

TABLE 6.5   Conventional generator characteristics

Thermal Unit	$P^{\min}$	$P^{\max}$	c_{nt}^u	c_{nt}^v	c_{nt}^q	c_{nt}^{rU}	c_{nt}^{rD}	MU_n	MD_n
1	50	260	100	100	8	5	5	7	7
2	50	200	85	85	10	6	6	6	6
3	30	150	70	70	12	5	5	5	5
4	20	120	65	65	14	5	5	5	5
5	20	150	55	55	14	5	5	5	5
6	5	30	20	20	40	10	10	1	1
7	10	55	20	20	45	15	15	1	1
8	5	50	30	20	60	20	20	1	1
9	10	35	20	20	60	20	20	1	1
10	10	40	20	20	80	30	30	1	1

TABLE 6.6   Relative solar energy

T	1	2	3	4	5	6	7	8	9	10	11	12
γ_t	0	0	0	0	0.2	0.5	0.9	1	1	1	1	1

T	13	14	15	16	17	18	19	20	21	22	23	24
γ_t	1	1	1	1	0.9	0.5	0.2	0	0	0	0	0

TABLE 6.7   System load

T	1	2	3	4	5	6	7	8	9	10	11	12
D_t (MWh)	215	196	190	220	228	240	255	272	296	324	335	346

T	13	14	15	16	17	18	19	20	21	22	23	24
D_t (MWh)	320	296	265	243	275	316	332	352	330	308	270	243

6.5.1   As a Function of the Integration Level of Stochastic Units

In this section, we analyze the simulation results of the reserve valuation as a function of the integration level of stochastic units. First, we consider the risk-neutral stochastic reserve determination model (6.1) and its counterpart with fixed amounts of reserve, that is, model (6.1) with $\sum_{n\in\mathcal{N}}(r_{nt}^{U} + r_{nt}^{D}) \geq 15\%D_t$ for all $t \in \mathcal{T}$, and compare their optimal solutions at various values of INT% based on the average total costs obtained from an out-of-sample simulation. We summarize the simulation results in Table 6.8. From this table, we can observe that the average total costs of the stochastic optimal solutions with optimal and fixed reserves (see columns 2 and 3) decrease as the value of INT% increases from 5% to 30%. Meanwhile, we observe that the average total cost of the stochastic optimal solution with fixed reserves is always

TABLE 6.8   Out-of-sample performance comparison of the stochastic models with optimal and fixed reserves as a function of the integration level

	Out-of-Sample Performance (Stochastic Model)			
INT%	Opt. Res. ($)	Fix. Res. ($)	Comparison (%)	Res. Cost ($)
5	205,253	214,638	4.37	590
7.5	199,321	208,397	4.36	924
10	193,646	202,402	4.33	1276
12.5	188,245	196,692	4.29	1601
15	183,261	191,412	4.26	1919
17.5	178,506	186,370	4.22	2088
20	174,299	181,832	4.14	2365
22.5	170,240	177,460	4.07	2782
25	166,508	173,432	3.99	3035
27.5	162,982	169,642	3.93	3200
30	159,766	166,084	3.80	3538

TABLE 6.9   Out-of-sample performance comparison of the robust models with optimal and fixed reserves as a function of the integration level

| INT% | Out-of-Sample Performance (Robust Model) | | | | |
	Opt. Res. ($)	Fix. Res. ($)	Comparison (%)	Rob. vs. Sto. (%)	Res. Cost ($)
5	206,519	216,192	4.47	0.61	257
7.5	201,165	209,784	4.11	0.92	411
10	195,868	203,981	3.98	1.13	557
12.5	190,837	199,060	4.13	1.36	736
15	186,193	194,937	4.49	1.57	933
17.5	181,715	190,248	4.49	1.77	1147
20	177,343	183,909	3.57	1.72	1311
22.5	173,236	180,907	4.24	1.73	1522
25	169,772	177,161	4.17	1.92	1815
27.5	166,278	173,060	3.92	1.98	2007
30	163,539	169,512	3.52	2.31	2633

higher than the one with optimal reserves at all integration levels. Furthermore, as the value of INT% increases, the difference between the average total costs (see column 4) slightly decreases, and the difference is greater than 3.5% at all integration levels. This observation indicates that the value of reserve remains significant as the renewable energy is increasingly integrated into an energy system. Finally, we observe that the reserve total cost (see column 5) of the stochastic optimal solution increases proportionally as the value of INT% increases.

Second, we consider the adaptive robust reserve determination model (6.9) and its counterpart with fixed reserve, that is, model (6.9) with $\sum_{n \in \mathcal{N}} (r_{nt}^{U} + r_{nt}^{D}) \geq 15\% D_t$ for all $t \in \mathcal{T}$, and compare their optimal solutions at various values of INT% based on the average total costs obtained from an out-of-sample simulation. We summarize the simulation results in Table 6.9. From this table, we can observe that the average total costs of the robust optimal solutions with optimal and fixed reserves (see columns 2 and 3) decrease as the value of INT% increases from 5% to 30%. Meanwhile, we observe that the average total cost of the robust optimal solution with fixed reserves is always higher than the one with optimal reserves at all integration levels. Furthermore, as the value of INT% increases, the difference between the average total costs (see column 4) fluctuates and remains greater than 3.5%. This observation confirms that the value of reserve remains significant as the renewable energy is increasingly integrated into an energy system. In addition, we compare the performance of optimal reserves obtained from the stochastic model (6.1) and robust model (6.9) (see column 5). We observe that the average total cost of the robust optimal solution is higher than that of the stochastic optimal solution at all integration levels, and the difference increases as the value of INT% increases. This observation highlights the value of applying the stochastic model as the renewable energy is increasingly integrated into an energy system. Finally, we observe that the reserve total cost (see column 6) of the robust optimal solution increases proportionally as the value of INT% increases (except when INT% = 30). By comparing the average total costs and reserve total costs of the stochastic and robust optimal solutions (i.e., by comparing columns 2 and 5 in Table 6.8 with columns 2 and 6 in Table 6.9), we observe that the robust optimal solutions schedule larger generation amounts and smaller upward/downward reserve amounts than the stochastic optimal solutions do.

6.5.2 As a Function of the Variability of Stochastic Units

In this section, we analyze the simulation results of the reserve valuation as a function of the variability of stochastic units. First, we consider the risk-neutral stochastic reserve determination model (6.1) and its counterpart with fixed reserve, that is, model (6.1) with $\sum_{n\in\mathcal{N}}(r_{nt}^{U} + r_{nt}^{D}) \geq 15\%D_t$ for all $t \in \mathcal{T}$, and compare their optimal solutions at various values of VAR% based on the average total costs obtained from an out-of-sample simulation. We summarize the simulation results in Table 6.10. From this table, we can observe that the average total costs of the stochastic optimal solutions with optimal and fixed reserves (see columns 2 and 3) increase as the value of VAR% increases from 10% to 30%. This observation indicates that it becomes more costly for an energy system to accommodate renewable energy with higher variability. Meanwhile, we observe that the average total cost of the stochastic optimal solution with fixed reserves is always higher than the one with reserves at all variability levels. Furthermore, as the value of VAR% increases, the difference between the average total costs (see column 4) slightly decreases, and the difference is greater than 3% at all variability levels. This observation indicates that the value of reserve remains significant as the renewable energy become more fluctuating. Finally, we observe that the reserve total cost (see column 5) of the stochastic optimal solution increases proportionally as the value of VAR% increases.

TABLE 6.10 Out-of-sample performance comparison of the stochastic models with optimal and fixed reserves as a function of the variability

| VAR% | Out-of-Sample Performance (Stochastic Model) | | | |
	Opt. Res. ($)	Fix. Res. ($)	Comparison (%)	Res. Cost ($)
10	172,617	181,378	4.83	1181
12.5	173,003	181,464	4.66	1467
15	173,419	181,582	4.50	1748
17.5	173,855	181,720	4.33	2029
20	174,299	181,832	4.14	2365
22.5	174,699	181,918	3.97	2668
25	175,096	182,006	3.80	2971
27.5	175,496	182,115	3.63	3270
30	175,917	182,244	3.47	3548

TABLE 6.11 Out of sample performance comparison of the robust models with optimal and fixed reserves as a function of the variability

VAR%	Out-of-Sample Performance (Robust Model)				
	Opt. Res. ($)	Fix. Res. ($)	Comparison (%)	Rob. vs. Sto. (%)	Res. Cost ($)
10	174,098	181,974	4.33	0.85	656
12.5	174,989	182,978	4.37	1.13	819
15	175,741	183,729	4.35	1.32	983
17.5	176,542	183,186	3.63	1.52	1147
20	177,343	183,909	3.57	1.72	1311
22.5	178,163	185,184	3.79	1.94	1483
25	178,988	185,628	3.58	2.17	1615
27.5	179,937	188,080	4.33	2.47	1803
30	180,782	184,586	2.06	2.69	1967

Second, we consider the adaptive robust reserve determination model (6.9) and its counterpart with fixed reserve, that is, model (6.9) with $\sum_{n\in\mathcal{N}}(r_{nt}^{U} + r_{nt}^{D}) \geq 15\%D_t$ for all $t \in \mathcal{T}$, and compare their optimal solutions at various values of VAR% based on the average total costs obtained from an out-of-sample simulation. We summarize the simulation results in Table 6.11. From this table, we can observe that the average total costs of the stochastic optimal solutions with optimal and fixed reserves (see columns 2 and 3) increase (except in the last instance) as the value of VAR% increases from 10% to 30%. Similar to what we observe from Table 6.10, this observation indicates that it becomes more costly for an energy system to accommodate renewable energy with higher variability. Meanwhile, we observe that the average total cost of the robust optimal solution with fixed reserves is always higher than the one with optimal reserves at all variability levels. Furthermore, as the value of VAR% increases, the difference between the average total costs (see column 4) fluctuates and remains greater than 2% at all variability levels. Similar to what we observe from Table 6.10, this observation indicates that the value of reserve remains significant as the renewable energy become more fluctuating. In addition, we compare the performance of optimal reserves obtained from the stochastic model (6.1) and robust model (6.9) (see column 5). We observe that the average total cost of the robust optimal solution is higher than that of the stochastic optimal solution at all variability levels, and the difference increases as the value of VAR% increases. This observation highlights the value of applying stochastic model as the renewable energy become more fluctuating. Finally, we observe that the reserve total cost (see column 6) of the robust optimal solution increases proportionally as the value of VAR% increases. By comparing the average total costs and reserve total costs of the stochastic and robust optimal solutions (i.e., by comparing columns 2 and 5 in Table 6.10 with columns 2 and 6 in Table 6.11), we observe that the robust optimal solutions schedule larger generation amounts and smaller upward/downward reserve amounts than the stochastic optimal solutions do.

TABLE 6.12 CPU time used to solve instances at various integration levels

INT%	Stochastic Models		Robust Models	
	Optimal Reserve (s)	Fixed Reserve (s)	Optimal Reserve (s)	Fixed Reserve (s)
5	42.5	51.1	12.3	12.6
7.5	42.3	50.0	12.7	12.5
10	41.9	49.6	12.2	12.5
12.5	42.0	48.8	12.4	12.3
15	39.5	46.7	12.0	12.7
17.5	36.4	47.0	12.0	12.7
20	35.4	40.3	12.1	12.5
22.5	32.2	37.4	12.3	12.4
25	32.1	34.6	12.5	12.7
27.5	31.5	30.0	12.0	12.8
30	30.2	22.9	12.4	12.3

6.5.3 Computational Time

In this section, we report the CPU time needed to solve all instances described in Sections 6.5.1 and 6.5.2. All problem instances were solved by using CPLEX 12.2 at a PC with Intel Core 2.90 GHz and 8 GB memory. In Table 6.12, we display the CPU time (in seconds) used to solve the instances at various integration levels. We observe that all instances can be solved within one minute, and in general it takes a longer time to solve the stochastic models as compared to solving the robust models. Also, we observe that there is no significant difference in terms of solving time for stochastic/robust models with optimal or fixed reserves.

In Table 6.13, we display the CPU time (in seconds) used to solve the instances at various variability levels. Similar to the results presented in Table 6.12, we observe that all instances can be solved within one minute, and in general, it takes a longer

TABLE 6.13 CPU time used to solve instances at various variability levels

VAR%	Stochastic Models		Robust Models	
	Optimal Reserve (s)	Fixed Reserve (s)	Optimal Reserve (s)	Fixed Reserve (s)
10	35.0	36.6	12.4	12.2
12.5	34.6	38.9	12.6	12.7
15	34.9	39.0	12.3	12.2
17.5	35.2	40.2	11.9	12.4
20	35.7	41.1	12.1	12.2
22.5	36.0	45.0	12.3	12.5
25	37.3	42.6	12.7	12.3
27.5	36.9	47.7	12.0	12.2
30	39.2	48.6	12.8	13.0

TABLE 6.14   CPU time used to solve instances with various numbers of buses

# Buses	Stochastic Models		Robust Models	
	Optimal Reserve (s)	Fixed Reserves(s)	Optimal Reserve (s)	Fixed Reserves(s)
30	35.5	40.5	12.1	12.6
60	48.2	55.6	25.6	22.6
90	63.7	65.5	35.8	34.3
120	74.7	78.1	43.6	44.9
150	85.9	92.4	49.1	50.2
180	100.3	107.4	54.2	53.6

time to solve the stochastic models as compared to solving the robust models. Also, we observe that there is no significant difference in terms of solving time for stochastic/robust models with optimal or fixed reserves.

To demonstrate the scalability of the proposed algorithm, we expand the size of the energy system and report the CPU times used to solve instances with various numbers of buses in Table 6.14. More specifically, we increase the number of buses from 30 to 180, and add 10 more thermal units and 1 more solar unit each time the system is expanded by 30 buses. Meanwhile, the system load proportionally increases and the added 10 thermal units and 1 solar unit have the same characteristics as described in Tables 6.5 and 6.6. From Table 6.14, we can observe that the CPU seconds proportionally increase as the size of the energy system grows (see columns 2–5), and the increasing trend is approximately linear (for the stochastic models) or sub-linear (for the robust models). This demonstrates that the proposed algorithm is scalable for the energy system under consideration.

6.6   SUMMARY, CONCLUDING REMARKS, AND RESEARCH NEEDS

In this chapter, we provide a tutorial overview on how to optimally procure reserves using both stochastic programming and robust optimization models, and study the value of reserves in various parameter settings based on both models. Moreover, we analyze the way reserve level and reserve cost change with renewable integration and renewable variability. We remark that both models presented in this chapter relaxed several security constraints (e.g., ramping-rate restrictions, transmission line capacity limits, etc) to facilitate discovering the structural insights of the models and investigating the economical value of reserves. More detailed models that capture various types of reserves (e.g., 10-minute spinning, 10-minute non-spinning, non-synchronized, and frequency regulation) are much needed as renewable energy is increasingly integrated into electrical energy systems. Future work also includes investigating chance-constrained stochastic programming approaches on determining the reserve requirements, the impact of different uncertainty set definitions on reserve procurement and advanced solution algorithms, and the impact of using stochastic and robust formulations on the electricity prices in day-ahead and real-time markets.

APPENDIX 6.A NOMENCLATURE

Sets and Indices

$\mathcal{N} = \{1, \ldots, N\}$ Set of thermal units in the power grid.

n Index for thermal units.

$\mathcal{T} = \{1, \ldots, T\}$ Set of operating time intervals (e.g., hours).

t Index for operational time intervals.

Parameters

c_t^o Unit cost pertaining to the generation curtailment amount in interval t.

c_{nt}^q Unit cost pertaining to the actual generation amount of unit n in interval t.

c_{nt}^{rD} Unit cost pertaining to the downward reserve amount of unit n in interval t.

c_{nt}^{rU} Unit cost pertaining to the upward reserve amount of unit n in interval t.

c_t^s Unit cost pertaining to the load-shedding amount in interval t.

c_{nt}^u Unit cost pertaining to the start-up operation of unit n in interval t.

c_{nt}^v Unit cost pertaining to the shut-down operation of unit n in interval t.

c_t^w Unit cost pertaining to the amount of curtailed renewable energy in interval t.

c_{nt}^x Unit cost pertaining to the on/off status of unit n in interval t.

D_t System load in interval t.

MD_n Minimum-down time of unit n.

MU_n Minimum-up time of unit n.

P_n^{max} Maximal generation capacity of unit n.

P_n^{min} Minimal generation amount of unit n.

ξ_t Total amount of available renewable energy in interval t.

Decision Variables

g_{nt} Scheduled generation amount of unit n in interval t.

o_t Generation curtailment amount in interval t.

q_{nt} Actual generation amount of unit n in interval t.

r_{nt}^D Scheduled downward reserve amount of unit n in interval t.

r_{nt}^U Scheduled upward reserve amount of unit n in interval t.

s_t Load-shedding amount in interval t.

u_{nt} Binary decision variable to indicate the start-up operation of unit n at the beginning of interval t.

v_{nt} Binary decision variable to indicate the shut-down operation of unit n at the beginning of interval t.

w_t Curtailed amount of renewable energy in interval t.

x_{nt} Binary decision variable to indicate the on/off status of unit n in interval t.

y_t Utilized amount of renewable energy in interval t.

REFERENCES

[1] "PJM Interconnection LLC." Available at http://www.pjm.com/Default.aspx, 2014.

[2] Federal Energy Regulatory Commission (FERC), "Promotion of wholesale competition through open access non-discriminatory transmission services by public utilities and recovery of stranded costs by public utilities and transmitting utilities," April 1996. Order No. 888.

[3] Federal Energy Regulatory Commission (FERC), "Open access same-time information system (formerly, real-time information networks) and standards of control," April 1996. Order No. 889.

[4] Alberta Electric System Operation, "Phase Two Wind Integration." Available at http://www.aeso.ca/downloads/Phase_II_Wind_Integration_Recommendation_-_Final.pdf, 2012.

[5] California ISO, "Flexible ramping product." Available at http://www.caiso.com/informed/Pages/StakeholderProcesses/FlexibleRampingProduct.aspx, 2014.

[6] Electric Reliability Council of Texas (ERCOT), "Future Ancillary Services in ERCOT." Available at http://www.ercot.com/content/news/presentations/2014/ERCOT%20AS%20Concept%20Paper%20Version%201_0%20as%20of%209-27-13%201745.pdf, 2013. [Online Available].

[7] OMI-Polo Español S.A. (OMIE). Available at http://www.omel.es/en/inicio, 2014.

[8] Q. P. Zheng, J. Wang, and A. L. Liu, "Stochastic optimization for unit commitment - a review," *IEEE Transactions on Power Systems*, vol. 30, no. 4, pp. 1913–1924, 2015.

[9] S. W. Wallace and S.-E. Fleten, "Stochastic programming models in energy," *Handbooks in Operations Research and Management Science*, vol. 10, pp. 637–677, 2003.

[10] A. Tuohy, P. Meibom, E. Denny, and M. O'Malley, "Unit commitment for systems with significant wind penetration," *IEEE Transactions on Power Systems*, vol. 24, no. 2, pp. 592–601, 2009.

[11] J. M. Morales, A. J. Conejo, and J. Pérez-Ruiz, "Economic valuation of reserves in power systems with high penetration of wind power," *IEEE Transactions on Power Systems*, vol. 24, pp. 900–910, May 2009.

[12] F. Bouffard, F. D. Galiana, and A. J. Conejo, "Market-clearing with stochastic security - part i: formulation," *IEEE Transactions on Power Systems*, vol. 20, no. 4, pp. 1818–1826, 2005.

[13] F. Bouffard, F. D. Galiana, and A. J. Conejo, "Market-clearing with stochastic security - part ii: case studies," *IEEE Transactions on Power Systems*, vol. 20, no. 4, pp. 1827–1835, 2005.

[14] L. Wu, M. Shahidehpour, and T. Li, "Stochastic security-constrained unit commitment," *IEEE Transactions on Power Systems*, vol. 22, no. 2, pp. 800–811, 2007.

[15] Q. Wang, Y. Guan, and J. Wang, "A chance-constrained two-stage stochastic program for unit commitment with uncertain wind power output," *IEEE Transactions on Power Systems*, vol. 27, no. 1, pp. 206–215, 2012.

[16] D. G. Luenberger and Y. Ye, *Linear and Nonlinear Programming*. Springer, 2008.

[17] A. J. Conejo, E. Castillo, R. Mínguez, and R. García-Bertrand, *Decomposition Techniques in Mathematical Programming. Engineering and Science Applications*. Springer, 2006.

[18] L. Baringo and A. J. Conejo, "Offering strategy via robust optimization," *IEEE Transactions on Power Systems*, vol. 26, no. 3, pp. 1418–1425, 2011.

[19] R. Jiang, J. Wang, and Y. Guan, "Robust unit commitment with wind power and pumped storage hydro," *IEEE Transactions on Power Systems*, vol. 27, no. 2, pp. 800–810, 2012.

[20] D. Bertsimas, E. Litvinov, X. A. Sun, J. Zhao, and T. Zheng, "Adaptive robust optimization for the security constrained unit commitment problem," *IEEE Transactions on Power Systems*, vol. 28, no. 1, pp. 52–63, 2013.

[21] R. Jiang, M. Zhang, G. Li, and Y. Guan, "Two-stage network constrained robust unit commitment problem," *European Journal of Operational Research*, vol. 234, no. 3, pp. 751–762, 2014.

[22] A. Thiele, T. Terry, and M. Epelman, "Robust linear optimization with recourse," technical report, Lehigh University, 2009.

[23] L. Zhao and B. Zeng, "Robust unit commitment problem with demand response and wind energy," in *IEEE Power and Energy Society General Meeting*, pp. 1–8, IEEE, 2012.

[24] G. P. McCormick, "Computability of global solutions to factorable nonconvex programs: part i - convex underestimating problems," *Mathematical Programming*, vol. 10, no. 1, pp. 147–175, 1976.

PART IV

HARNESS
TRANSMISSION
FLEXIBILITY

IMPROVED MARKET EFFICIENCY VIA TRANSMISSION SWITCHING AND OUTAGE EVALUATION IN SYSTEM OPERATIONS

Kwok W. Cheung and Jun Wu

IN **THIS CHAPTER**, we investigate the improvement of market efficiency via transmission switching and outage evaluation in system operations. Section 7.1 discusses the background of and challenges in power system operations. Section 7.2 reviews the basic dispatch model for market clearing. Section 7.3 describes the state-of-the-art process of outage coordination and a measure of economic evaluation of transmission outages. Section 7.4 explores the concept of optimal transmission switching and the enhancement of the dispatch model to incorporate transmission switching within the context of a security-constrained economic dispatch algorithm. Section 7.5 proposes a new criterion for selecting candidates for transmission-line switching based on production cost (PC). In order to achieve a better market surplus within a reasonable time frame, a new method of identifying a limited set of transmission line candidates for switching that is combined with other selection criteria is proposed. Numerical examples are given to illustrate the effectiveness of the proposed method using a practical large-scale power system in Section 7.6. Finally, some of the market design and implementation challenges are considered in Section 7.7.

7.1 BACKGROUND

The restructured electric power industry has brought new challenges for the secure and efficient operation of stressed power systems. In North America, almost all regional transmission organizations (RTOs), such as PJM, Mid-continent ISO, or ISO New England, are fundamentally reliant on the wholesale market mechanism to optimally dispatch energy and ancillary services of generation resources in order to reliably serve the load in the large geographical region [1–3].

Power Grid Operation in a Market Environment: Economic Efficiency and Risk Mitigation, First Edition.
Edited by Hong Chen.

With increasing threats of major blackouts and persistent incentives to improve revenue performance, transmission system operators face daunting challenges. Renewable generation such as extensive penetration of the wind power, demand resources, distributed energy resources, and emission management is adding significant burdens on system operation to provide transmission services with unparalleled agility and resiliency [4]. In conjunction with technologies, such as high-performance communications, phasor measurement units (PMUs), and power electronic devices that offer potential benefits to grid operators, a more controllable and flexible transmission system can help operators. With early warning applications, operators can improve their awareness of the overall grid security situation and devise plans for preemptive defense and/or corrective remedies [5].

Traditionally, power system topology is considered "fixed" and non-dispatchable in the sense that transmission outage and switching is not part of the market optimization or any economic consideration. Most outage and transmission switching analysis is solely based on system reliability criteria without much consideration of market efficiency. Today, the control of a transmission element's state is via an EMS/SCADA interface to open or close a circuit breaker. To avoid transmission congestion or improve voltage profiles, system operators are often required to change the topology of the network based on their own experience or a set of heuristic operating procedures on an *ad hoc* basis. It is desirable to have an analysis tool to assess the economic impacts of transmission outages and switching and to drive improvement of market efficiency.

In recent years, energy systems (whether in developed or in emerging economies) have been undergoing changes due to the challenges imposed by the smart grid. Optimal transmission switching [6–9] has received some attention and seems to be a viable way to leverage grid controllability for enhancing system performance in short-term system operation [10, 11] as well as long-term system planning [12]. It is obvious that transmission systems could become smarter and more efficient when the control of network topology can be factored into the overall dispatch process of generation and transmission resources, taking both system reliability and economics into consideration in a systematic fashion. However, more analysis and investigation to assess practical implications such as revenue adequacy, stability concerns, and computational performance is required before putting optimal transmission switching into practice.

7.2 BASIC DISPATCH MODEL FOR MARKET CLEARING

A typical market system in North America supports a series of business functions for market and system operations [13]. Depending on the operational environment, the system can be configured to include one or more of the following business processes (Figure 7.1):

- Day-ahead market (DAM) process—provides functions for day-ahead bid data submission, market clearing, and market solution publishing.

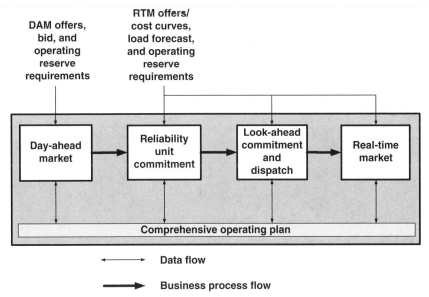

Figure 7.1   Business processes of a typical market system.

- Reliability unit commitment (RUC) process—provides system operators with a set of tools to revise the day-ahead unit commitment schedule as necessary in order to ensure that the forecasted load and operating reserve requirements will be met and that the transmission system is reliable and secure.

- Look-ahead commitment and dispatch (LACD) process—provides a forward-looking view of system operating conditions and offers start-up/shut-down recommendations of fast-start resources to operators.

- Real-time market (RTM) process—provides market-clearing functions to balance generation and load and meet reserve requirements based on actual real-time system operating conditions. The RTM process computes ex ante pricing and provides the dispatch signals in the form of either megawatts or price back to the market participants.

The core application of the market system is basically a centralized market clearing engine solving a unit commitment, scheduling, and dispatch problem. The problem itself is formulated as a mixed integer programming (MIP) problem [13]. A compact form of such a problem can be described as follows:

$$\min \sum_{g,t} (u_{gt}\chi_{gt}(p_{gt}) + \varsigma_{gt}(u_{g(t-1)}, u_{gt})) \tag{7.1}$$

subject to

$$(\lambda_t) \qquad \sum_g p_{mgt} = l_t + p_t^{loss}, \quad \forall t \tag{7.2}$$

$$\sum_g r_{gt} \geq \underline{r}_t, \quad \forall t \tag{7.3}$$

$$u_{gt}\underline{p}_{gt} \leq p_{mgt} \leq u_{gt}\bar{p}_{gt}, \quad \forall g, t \tag{7.4}$$

$$p_{gt} + r_{gt} \leq u_{gt}\bar{p}_{gt}, \quad \forall g, t \tag{7.5}$$

$$0 \leq r_{gt} \leq u_{gt}\bar{r}_{gt}, \quad \forall g, t \tag{7.6}$$

$$(\mu_{kt}) \quad \underline{f}_{kt} \leq f_{kt} \leq \bar{f}_{kt}, \quad \forall k, t \tag{7.7a}$$

$$f_{kt} = B_{mn}(\theta_{mt} - \theta_{nt}), \quad \forall k, t \tag{7.7b}$$

$$p_{gt} - l_{mt} - p_{mt}^{loss} = \sum_{k \in line_m^{fr}} f_{kt} - \sum_{k \in line_m^{to}} f_{kt}, \quad \forall m, t \tag{7.8a}$$

$$l_t = \sum_m l_{mt}, \quad \forall t \tag{7.8b}$$

$$p_t^{loss} = \sum_m p_{mt}^{loss}, \quad \forall t \tag{7.8c}$$

$$\underline{\theta} \leq \theta_{mt} \leq \bar{\theta}, \quad \forall m, t \tag{7.9}$$

$$(\mathbf{u}, \mathbf{p}, \mathbf{r}) \in \Gamma \tag{7.10}$$

See Appendix 7.A for the summary of nomenclature. The objective (7.1) is to minimize the production plus start-up costs. The minimization is subject to many constraints including supply and demand constraints (7.2), capacity constraints (7.4), (7.5), and (7.6), transmission constraints (7.7) and (7.8), and spinning reserve requirements (7.3). It is important to note that the transmission flow can also be expressed as a linear function of bus net injections in the pre-contingent state.

$$f_{kt} = \sum_g a_{kgt}p_{gt}, \quad \forall k, t \tag{7.11}$$

Most RTOs' real-time operation requires the $n - 1$ security protection, and therefore the transmission flow presented here can be either pre-contingent or post-contingent line or interface flow. Other constraints, such as minimum up and down times constraints, ramp constraints, and operating and regulating reserve requirements, are considered as a part of (7.10).

After the unit commitment problem is solved, the integer variables u_{gt} will be frozen, and the optimal solution for the scheduling and dispatch problem will be reduced to a linear programming (LP) problem in which the market clearing quantities, constraint shadow prices (Lagrange multipliers), and market clearing prices by location can be determined. The market clearing prices by location, called locational marginal prices, are by-products of the optimization solution. Locational marginal prices for energy can be written as

$$
\text{LMP}_{gt} = \lambda_t - \lambda_t \frac{\partial p_t^{\text{loss}}}{\partial p_{gt}} - \sum_k a_{kgt} \mu_{kt} \tag{7.12}
$$

The three terms in the above LMP equation could be interpreted as the three components of LMP, namely energy, loss, and congestion, respectively. These locational price results give a precise representation of the cause–effect relationship that is consistent with grid reliability management. Note that the formulation of joint optimization between energy and ancillary services is ignored here for the sake of simplicity.

7.3 ECONOMIC EVALUATION OF TRANSMISSION OUTAGE

Transmission outages are critical inputs to the market system of RTOs. There are three major processes for transmission outage coordination:

1. Outage submission process: Local control center submits an outage application to the RTO.

2. Outage review process: RTO's outage coordinator reviews the outage based on system reliability (reliability study) and economics (economic impact study) and makes a decision on the approval, denial, and reposition of the outage.

3. Outage communication process: Outage status is communicated back to the local control center and market participants. A negotiation effort may start to reposition the requested outage.

Economic evaluation of transmission outages [14] comes into play in the outage review process, particularly in the economic impact analysis of transmission outages for both long term (weeks to months ahead) and short term (a few days ahead).

A measure called outage-caused congestion exposure (OCCE) is defined to reflect the impact of a transmission outage on market efficiency. While OCCE in dollars is normally positive, it can be a negative quantity if the transmission outage actually reduces congestion. There are two types of OCCE: existing and incremental. Existing OCCE is defined as the PC of a study period with a previously *approved* transmission outage minus the PC of a study without any transmission outages. Incremental OCCE is defined as the PC of a study period with a *proposed* transmission outage minus the PC of a study without the transmission outage. The PC is the evaluation of objective function (7.1).

At a high level, the intent of the economic analysis of transmission outage requests is to find opportunities for improved placement of transmission outages which will result in lower costs to the market. In general, there is a long-term outage coordination process looking ahead a month to a year and a short-term outage coordination process just a few days to a couple of weeks ahead. Two analysis subprocesses are part of the long-term process while one analysis subprocess is part of the short-term process. These subprocesses are summarized below:

- Individual evaluation: Every outage that passes the reliability analysis receives an individual evaluation to determine the incremental OCCE impact of that outage. A study is performed containing cases with interim-approved transmission outages as well as cases with interim-approved and proposed transmission outages. The increase in PC is used to determine the incremental OCCE for that proposed outage. For the majority of transmission outages, the incremental OCCE will be below a specified threshold. For the transmission outages meeting the threshold, the transmission outage will automatically be approved and given the state of interim approval. If an individual evaluation determines that the incremental OCCE exceeds the dollar threshold, the outage coordinator will then perform a scenario analysis on that proposed transmission outage.

- Scenario analysis: Because a proposed transmission outage did not meet the individual analysis threshold criteria, the transmission outage request requires a thorough economic evaluation. The outage coordinator will analyze multiple scenarios to determine the impact of placing the transmission outage in various time periods. If the outage coordinator determines that there are time periods with lower incremental OCCE for the transmission outage, the outage coordinator will then negotiate with the transmission owner to see if the transmission outage can be rescheduled to a more favorable time period. The results of the scenario analysis will provide the outage coordinator with the scenario information needed to negotiate a more favorable outcome. Figure 7.2 summarizes the overall long-term outage coordination process.

- Short-term economic assessment: Most transmission outage requests are evaluated weeks or months in advance when critical assumptions such as weather, load, and unplanned transmission outages have high degrees of uncertainty. As the outage window becomes within a few days of the transmission outage, it is essential to reevaluate all transmission outage requests utilizing the latest assumptions, including network configurations, load forecasts, short-term generation and transmission outage submittals, and unplanned generation and transmission outages. The outage coordinator is concerned with occurrences where scheduled generation and transmission outages may have substantial impacts on prices or revenue adequacy. By providing an assessment of potential market inefficiencies a few days before the actual outage, system and market operations are afforded an opportunity to evaluate actions that could alleviate the economic exposure—actions such as committing long run-time units, canceling the outage, or sending out notifications.

Economic evaluation of transmission outages is only the first step toward market efficiency improvement. Optimal scheduling of outages in operations planning and optimal transmission switching in real-time dispatch are both potential avenues to further improve economic efficiency. The following section will briefly discuss the latter.

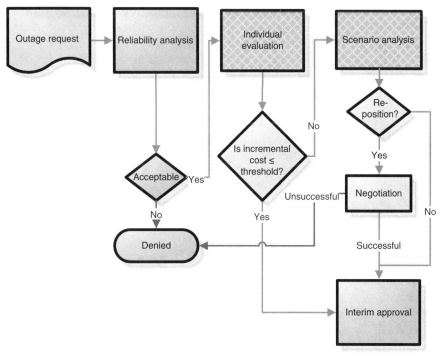

Figure 7.2   Long-term outage coordination process.

7.4   OPTIMAL TRANSMISSION SWITCHING

Transmission elements are traditionally treated as nondispatchable asset in the network. Although the state of a transmission element can manually be controlled by system operators, the state is typically not a decision variable of the optimization problem. Co-optimizing transmission topology and generation dispatch could be a viable way to further maximize the market surplus and improve economic efficiency. It is not hard to imagine that as load patterns change, the optimal topology can change as well. Having an active and dynamic topology network adapting the uncertainty of future system conditions makes transmission switching beneficial.

Transmission switching formulations have been previously presented in [6, 7]. The basic dispatch model (7.1)–(7.10) can be modified to create an optimal transmission switching (OTS) model by replacing (7.7) and (7.8) with the following equations:

$$\underline{f}_{kt} z_{kt} \leq f_{kt} \leq \bar{f}_{kt} z_{kt}, \quad \forall k, t \tag{7.13}$$

$$B_{mn}(\theta_{mt} - \theta_{nt}) - f_{kt} + (1 - z_{kt})M_k \geq 0, \quad \forall k, t \tag{7.14a}$$

$$B_{mn}(\theta_{mt} - \theta_{nt}) - f_{kt} - (1 - z_{kt})M_k \geq 0, \quad \forall k, t \tag{7.14b}$$

Note that M_k listed in (7.14a) and (7.14b) is an arbitrarily large number. z_{kt} is the binary variable representing the state of the transmission element; a value of 1 reflects a closed status and a value of 0 reflects an open status. When the binary variable z_{kt} is 1, the value of M_k is irrelevant. When the binary variable z_{kt} is 0, the value of M_k should be big enough to ensure that (7.14a) and (7.14b) are satisfied regardless of the difference in the bus angles. For all practical purposes, M_k can be set to $B_{mn}\left(\bar{\theta} - \underline{\theta}\right)$.

Using a simulation tool, one could compare the PC (7.1) of the basic dispatch with that of the OTS model and also assess the benefit if deploying the transmission switching model. Although the changes required in the mathematical formulation are seemingly quite simple, the computational performance impact might be an issue and deserves more studies. Due to performance concerns, most state-of-the-art real-time dispatch engines for market clearing in North American wholesale electricity markets are currently based on a power transfer distribution factor (PTDF) approach to model transmission constraints instead of using a DC power flow model inside the formulation. As indicated in (7.13) and (7.14), a DC power flow is required for the modeling of OTS.

For a large-scale power system with roughly 37,000 buses, 47,000 branches, and 1700 generating units, test cases of security-constrained economic dispatch are run. The dispatch engine is based on a single-period LP problem (without transmission switching), which is discussed in Section 7.2. Table 7.1 illustrates the performance comparison of the dispatch engine with and without the DC power flow model under various system conditions. One can observe that the performance with the DC power flow model is significantly slower than the one without the DC power flow model. The difference is mainly due to the introduction of branch flow constraints in the DC power flow model.

Besides performance degradation due to the power flow model, the number of integer variables being introduced could be significant and hence severely impact the computational performance. One way to limit the number of integer variables is to introduce the following constraints:

$$\sum_k (1 - z_{kt}) \leq J, \quad \forall k, t \tag{7.15}$$

where J denotes the maximum number of open lines. For the same example of a large-scale test system with more than 47,000 branches, there are mathematically speaking more than $2^{47,000}$ potential combinations of transmission topology. However, it is

TABLE 7.1   DC power flow performance comparison

	Solver Solution Time (s)	
	LP without DC model	LP with DC model
Condition 1	2.34	272.83
Condition 2	12.55	459.50
Condition 3	14.04	551.95

TABLE 7.2   Solution times vary with OTSBranch and *J*

		Solver Solution Time (s)
LP w/o OTS but w/ DC Power Flow		302
MIP w/ OTS	OTSBranch = 18; $J = 3$	338
	OTSBranch = 18; $J = 7$	378
	OTSBranch = 18; $J = 18$	378
	OTSBranch = 102; $J = 18$	58,189
	OTSBranch = 340; $J = 5$	2,444
	OTSBranch = 691; $J = 5$	3,138
	OTSBranch = 1561; $J = 5$	4,686
	OTSBranch = 1647; $J = 5$	7,202

important to note that many combinations might not make any engineering sense and could be eliminated heuristically. Even limiting *J* to a small number, the optimization problem is still very computationally challenging, if not impractical, to solve. To be able to practically solve the MIP model for such a large-scale system, we further limit the set of switchable transmission lines so that z_{kt} of the non-switchable subset are frozen. Table 7.2 illustrates the computational time for various values of *J* and the size of the switchable set. By controlling both the size of the switchable set (OTSBranch) and the value of *J* below a certain value, it is realistic to solve the problem in a reasonable time frame. This is similar to the problem without the augmented formulation of OTS but with the DC power flow model. Although the solution is suboptimal due to the additional constraints of OTSBranch limits and (7.15), recommended transmission switching does improve the market efficiency. Figure 7.3 shows a small 5-bus test case [14] with both OTSBranch and *J* set to 10. OTSOB and LPOB illustrate the

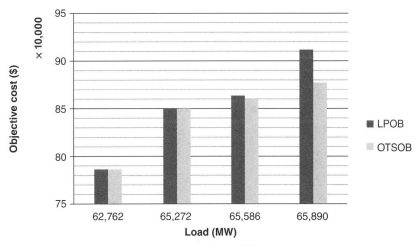

Figure 7.3   Comparison of objective cost due to OTS.

objective costs for the model with and without OTS model, respectively. For the OTS case, two lines are switched off for all load conditions. Note that total savings due to OTS becomes more apparent as the system is getting more stressed.

7.5 SELECTION OF CANDIDATE TRANSMISSION LINES FOR SWITCHING AND IMPLEMENTATION OF OTS

With the introduction of z_{kt} in the OTS model, transmission switching adds substantial computational complexity to the market clearing engine. To be able to solve the MIP model for large-scale systems, transmission lines are divided into two sets. One is a set of switchable transmission lines which are associated with decision integer variables as in (7.7)–(7.10). The other is a set of transmission lines in which status (on/off) is predetermined before running optimization of market clearing.

The proper size of switchable transmission lines and the value of J in (7.10) are critical parameters that affect solution time. These parameters might vary from network to network and case by case. Our previous research [15, 16] has shown that solving OTS could be done within the required time in large electricity market systems by selecting a proper size of candidate transmission lines. However, it remains a challenge to select a proper set of candidate transmission lines to get better objective savings among thousands of transmission lines for a given large-scale power system. The following criteria are proposed to be the selection criteria of good candidate transmission lines for switching [16].

7.5.1 Selection Criteria

7.5.1.1 Limited Violations of Transmission Lines
For a given base case, if transmission lines violate their limits, the violation penalties will appear in the cost of the objective function. Switching off the violated transmission lines does not cause any new violations and will eliminate the high penalty costs introduced by those violated transmission lines. Hence, the violated transmission lines could be good candidates for switchable transmission lines.

7.5.1.2 Congestion Rents of Transmission Lines
Congestion rent (CR) [7] is defined as the difference in shadow prices between the "*from*" and "*to*" side buses of transmission line k, multiplied by the flow on the line:

$$CR = f_{kt}(\lambda_{mt} - \lambda_{nt}) \tag{7.16}$$

By introducing a fictitious variable of "fraction of line out of service," Fuller et al. [8] suggested using (7.16) to rank transmission lines for switching. Due to complementary slackness, the shadow price of a transmission line is zero if there are no transmission constraints binding, negative if there are transmission constraints binding at upper limits, and positive if there are transmission constraints binding at lower limits. The lines are congested when there are shadow price differences between the "*from*" and "*to*" sides of the line of interest. Switching off these transmission lines would possibly reduce the congestion costs.

7.5.1.3 Production Costs Associated with Transmission Lines

As mentioned in Section 7.2 the objective of OTS for scheduling and dispatch is to minimize PC while satisfying transmission constraints and other operating constraints. The PC could be increased due to transmission congestion. In those cases, some higher-cost generation is dispatched in favor of lower-cost generation that would otherwise be used in the absence of transmission constraints.

Assuming solving for a DC power flow solution, the transmission flow is expressed as follows:

$$f_{k(m,n)t} = B_{mn}(\theta_m - \theta_n) \tag{7.17}$$

Suppose ΔF is the vector of changes of transmission line flow $f_{k(m,n)t}$, ΔF can be written as

$$\Delta F = A \Delta \theta \tag{7.18}$$

where matrix A has the following structure:

$$A = \begin{bmatrix} & \text{col. } m & & \text{col. } n & \\ .. & .. & .. & .. & .. \\ 0.. & B_{mn} & 0.. & -B_{mn} & 0.. \\ .. & .. & .. & .. & .. \end{bmatrix}$$

Corresponding to a transmission line, each row of A contains only two nonzero elements of positive and negative values of the line susceptance in columns m and n, respectively. Pre-multiplying both sides of equation (7.18) by A', the transpose of A,

$$A'\Delta F = A'A\Delta\theta = \Pi\Delta\theta \tag{7.19}$$

where

$$\Pi_{ij} = \begin{cases} \sum B_{ii}^2, & i = j \\ -\sum B_{ij}^2, & i \neq j \end{cases} \tag{7.20}$$

i, j are bus nodes and the left-hand side of (7.19) becomes

$$A'\Delta F = \begin{bmatrix} \sum\limits_{\substack{i \\ k\in\text{line}_i^{\text{fr}}}} B_{1i}\Delta f_{kt} - \sum\limits_{\substack{i \\ k\in\text{line}_i^{\text{to}}}} B_{1i} f_{kt} \\ \vdots \\ \sum\limits_{\substack{m \\ k\in\text{line}_m^{\text{fr}}}} B_{mi}\Delta f_{kt} - \sum\limits_{\substack{m \\ k\in\text{line}_m^{\text{to}}}} B_{mi} f_{kt} \\ \vdots \\ \sum\limits_{\substack{n \\ k\in\text{line}_n^{\text{fr}}}} B_{ni}\Delta f_{kt} - \sum\limits_{\substack{n \\ k\in\text{line}_n^{\text{to}}}} B_{ni} f_{kt} \end{bmatrix} \tag{7.21}$$

Equating (7.19) and (7.21), $\Delta\theta_i$ can be calculated as

$$\Delta\theta_{it} = \sum_k B_{mn}\left(\prod_{mi}^{-1} - \prod_{ni}^{-1}\right)\Delta f_{k(m,n)t} = \sum_k \beta_{ik}\Delta f_{k(m,n)t} \tag{7.22}$$

where

$$\beta_{ik} = B_{mn}\left(\prod_{mi}^{-1} - \prod_{ni}^{-1}\right) \tag{7.23}$$

can be interpreted as the sensitivities of the bus angle of bus i with respect to the flow of transmission line k. In other words,

$$\frac{\partial\theta_{it}}{\partial f_{kt}} = \beta_{ik} \tag{7.24}$$

Let the PC be

$$PC = \sum c_{gt} p_{gt} \tag{7.25}$$

where c_{gt} is the energy offer price of generator g at time t. The sensitivity of PC with respect to transmission line flow f_{kt} is

$$\frac{\partial PC}{\partial f_{kt}} = \sum_g \frac{\partial PC}{\partial p_{gt}}\frac{\partial p_{gt}}{\partial f_{kt}} \tag{7.26}$$

Assuming the system has negligible losses and no dispatchable loads, and using (7.24), we can derive

$$\frac{\partial p_{gt}}{\partial f_{kt}} = \sum_i \frac{\partial p_{gt}}{\partial\theta_{it}}\frac{\partial\theta_{it}}{\partial f_{kt}}$$
$$= B_{ii}\beta_{ik} - \sum_{h\neq i} B_{hi}\beta_{ik} \tag{7.27}$$

As a result,

$$\frac{\partial PC}{\partial f_{kt}} = \sum_g \frac{\partial PC}{\partial p_{gt}}\frac{\partial p_{gt}}{\partial f_{kt}}$$
$$= \sum_g c_{gt}\left(B_{ii}\beta_{ik} - \sum_{h\neq i} B_{hi}\beta_{ik}\right) \tag{7.28}$$

Hence, the PC associated with the transmission line k, PC_k, can be defined as

$$PC_k = \left(\sum_g c_{gt}\left(B_{ii}\beta_{ik} - \sum_{h\neq i} B_{hi}\beta_{ik}\right)\right)f_k \tag{7.29}$$

which can be used as another selection criterion for transmission switching candidates.

TABLE 7.3   Solution time varies with OTSBr and *J*

OTSBr	*J*	# of Lines Opened	Solution Time (s)
14	10	1	351
18	8	4	378
18	18	4	378
27	5	4	516
74	5	5	920
340	5	1	2444

7.5.2 Selection Method for Transmission Switching Candidates

We propose a selection method for transmission switching candidates that uses the combined set of selection criteria described above. Selection criteria are heuristically combined together to filter the set of transmission lines and determine the switching candidates. The proposed approach turns out to give better results compared to any single criterion applied alone. The rest of this section will describe the proposed selection method.

The violated transmission lines, if any, will first be selected as switching candidates since they usually cause large penalty costs in the objective function. However, this criterion is usually not satisfied under normal system conditions. Instead, binding transmission constraints are more likely to be observed.

Both CR and PC_k can be used as ranking measures for effective transmission switching candidates. Basically, the higher their values, the higher the ranking they will be as transmission switching candidates. However, it is important to note that neither CR nor PC_k is a perfect measure from an economic perspective. Our experiences show that combining PC_k and CR (PCCR) seems to be a better indicator for the selection of switching candidates compared to the selection method based on CR alone.

For all practical purposes, the OTS problem could be solved by selecting one candidate line at a time [6, 8] or limiting the maximum number of open lines. In either case, a proper size of the candidate set of transmission lines [9] is crucial to making the problem tractable.

Table 7.3 illustrates the OTS solver solution time for differently sized candidate transmission lines (OTSBr) and the different maximum number of open lines *J*. The practical number of switchable transmission lines is within 20 and the maximum number of open lines *J* is about 10. Tractable values of OTSBr and *J* could vary with the size of the power network. Fortunately, based on our practical experience so far, these parameters do not usually vary much for a given power network and can be found from offline studies.

7.5.3 Implementation of Optimal Transmission Switching

The implementation of solving OTS using the PCCR for OTS candidates is described by the following procedure:

Step 1: Calculate sensitivities β_{ik} in the pre-contingent state.

Step 2: Solve the model without OTS.

Step 3: Add violated transmission lines, if any, to the set of transmission line candidates.

Step 4: If the number of candidate transmission lines does not exceed the parameter of OTSBr, calculate CR and PC_k for each of the rest of the transmission lines according to equations (7.16) and (7.29), respectively.

Step 5: Rank transmission lines by CR, filter out nonpositive PC_k, and add the top ranking transmission lines to the set of transmission line candidates until the OTSBr limit is reached.

Step 6: Solve the OTS problem using the MIP formulation of (7.1)–(7.13) and output the results.

Figure 7.4 illustrates the procedure for selecting transmission switching candidates.

For the sake of simplicity, $N-1$ security is not considered here. However, for all practical purposes, it is expected that the OTS solution of generation dispatch and the recommended topology switching will be validated by $N-1$ contingency analysis and a rerun of OTS with additional enforcement of contingency constraints may deem necessary.

7.6 TEST CASES

We have tested OTS using a 5-bus test case and also a large system with more than 37,000 buses, 47,000 branches, and 1700 generating units. The 5-bus test case is a simple example to illustrate the basic concept of OTS. For the large system, multiple case studies were performed under system conditions of different stress and transmission congestion levels.

7.6.1 5-Bus Test Case

Figure 7.5 is the one-line diagram for a 5-bus test case. There are seven transmission lines, three generating units, and three loads in this network. The total system load is 950 MW distributed on buses B, C, and D as 216.67, 316.67, and 416.66 MW, respectively. All generating units are dispatchable between the high and low operating limits (HOL/LOL). HOL/LOL and energy offer prices of generating units are shown in Table 7.4. Line characteristics are shown in Table 7.5.

For the sake of simplicity, there is no reserve requirement in this 5-bus case and energy is the only commodity. Assuming all 7 lines are switchable and J in (7.15) is set to 7, lines B–C and C–D are recommended to be switched off by the OTS. Tables 7.6, 7.7, and 7.8 show cleared megawatts, transmission line flows, and objective costs, respectively, for cases with and without OTS. The total energy cost saving due to OTS is $776/h, which is about 1.55% of the total energy cost. If J is set to 1, line C–D will

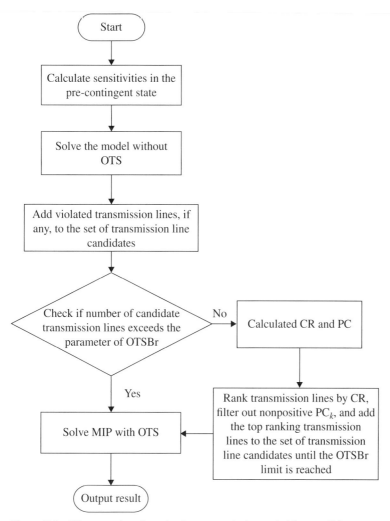

Figure 7.4   The procedure for selecting transmission switching candidates.

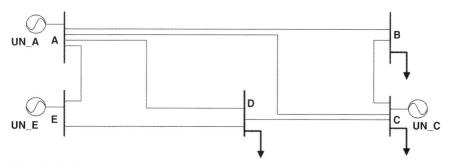

Figure 7.5   One-line diagram of a 5-bus system.

TABLE 7.4 Generating unit characteristics and energy offers

Generator	HOL (MW)	LOL (MW)	Price ($/MW)
UN_A	599	250	185
UN_C	520	0	24
UN_E	640	180	2.5

TABLE 7.5 Transmission line characteristics

Line	From Bus	To Bus	Admittance (p.u.)	Flow Limit (MW)
A–B	A	B	−28.1	310
A–D	A	D	−30.4	530
A–E	A	E	−6.4	180
A–C	A	C	−6.4	540
B–C	B	C	−10.8	400
C–D	C	D	−29.7	400
D–E	D	E	−29.7	540

TABLE 7.6 Comparison of cleared megawatts with and without OTS

Generator	Dispatch Megawatts (w/o OTS)	Dispatch Megawatts (w/ OTS)
UN_A	250	250
UN_C	100.6	64.5
UN_E	599.4	635.5

TABLE 7.7 Comparison of line flows with and without OTS

Lines	Megawatts Flow (w/o OTS)	Megawatts Flow (w/ OTS)
D–E	−540	−540
A–B	−213.97	−216.67
A–D	48.26	−123.33
A–E	−59.38	−95.5
B–C	−2.69	0
A–C	47.14	252.17
C–D	−171.16	0

TABLE 7.8 Comparison of objective costs with and without OTS

	W/o OTS	W/ OTS
Objective ($/h)	50,163	49,387

be recommended to be switched off by the OTS. It is interesting to note that C–D happens to be the line which has the highest PC with nonnegative CR.

7.6.2 A Large System Test Case

The large system of interest is a real market system model of a large RTO in the Midwest with more than 37,000 buses, 47,000 branches, and 1700 generating units. With today's computer technology, it is practically impossible to solve the generic OTS/MIP model with more than $2^{47,000}$ potential transmission topologies within the required time frame of system operations. Defining a proper number of candidate transmission lines and a proper value of J seems to be a tractable solution for such a large-scale system. The proposed methodology to reduce the number of candidate transmission lines is applied. Multiple scenarios of different system conditions are studied.

7.6.2.1 Scenario 1: With Violated Transmission Lines
There are 18 violated transmission lines in this case. Choosing all of them as candidate transmission lines, 7 lines are switched off by the OTS and the objective cost has changed significantly from \$3,384,616/h to \$2,749,157/h, while solver solution times are almost the same.

7.6.2.2 Scenario 2: With Binding Transmission Lines
There are about 44 binding transmission constraints and 370 dispatchable genera-tion units in this case. We have tested both the CR method and the proposed PCCR method. Using CR, there are 19 candidate transmission lines. After solving the OTS problem, four lines are open. The results are shown in Table 7.9 where switched-off lines are shown in bold.

On the other hand, using PCCR, there are 14 candidate transmission lines. After solving the OTS problem, seven lines are open, as shown in bold in Table 7.10. The generation cost changes with and without OTS for PCCR are depicted in Figure 7.6. PC savings are sorted in descending order by generation from left to right.

The objective comparison with and without OTS for CR and PCCR is shown in Table 7.11. The objective saving of CR is about \$11,474/h equal to about 1.3% saving; the objective saving of PCCR is about \$24,274/h equal to about 2.7% saving. The PCCR has \$12,800/h more savings than that of CR. This demonstrates that the PCCR method is more effective than the CR method.

7.6.2.3 Scenario 3: With Binding Transmission Lines at Valley Load
There are about 16 binding transmission constraints and the total load is 48,640 MW which is the valley load of a particular day. We apply the CR method and the proposed PCCR method to select 10 candidate transmission lines. The OTS results for CR and PCCR are shown in Table 7.12.

As a result of OTS, two lines are open using the CR method and five lines are open using the PCCR method. The objective saving of CR is about \$4520/h equal to about 0.58% of saving; the objective saving of PCCR is about \$5072/h equal to about 0.65% of saving. The PCCR has \$552/h more saving than that of CR.

TABLE 7.9 Candidates with CR

Transmission Lines	CR ($/h)
BURR_OAK_34513_A_LN	305,032.27
BURR_OAK_A_NO_2_XFMRAK_XF	239,439.35
PLYMOUT2_13821_NIPS_P_LN	187,571.80
EASTLAKE_11_EL_LM_1_LN	125,764.79
EAU_CLA_EAU_CKING34_1_1_LN	109,724.77
EWINAMAC_13832_A_LN	95,546.81
EASTLAKE_EL61_TR61_XF	81,929.61
EASTLAKE_EL62_TR62_XF	70,459.46
ARGENTA_ARGENTWIN_34_1_1_LN	70,111.24
ARGENTA_ARGENBATTL34_1_1_LN	65,547.23
MANSFLD2_MANS_HOYT_1_LN	**58,850.29**
AJ_MA_MO_AJ_MAMONRO34_1_1_LN	**58,639.49**
LLOYD_11_LY_KI_1_LN	55,949.56
PERRY_C_S_8_PY_EL_1_LN	49,271.90
EWINAMAC_13882_A_LN	**42,393.93**
LAMONT2_11_LM_LY_1_LN	37,299.71
MILAN3_MILANMAJES34_1_1_LN	37,116.47
SAMMIS_HIGHLND_SAMMIS_1_LN	**36,696.69**
HIGHLNDD_HGHLN_S_SPNG_1_LN	35,232.15

7.6.2.4 Scenario 4: With Binding Transmission Lines at Peak Load

There are about 28 binding transmission constraints and the total load is 57,213.5 MW which is the peak load of a particular day. Again 10 candidate transmission lines are selected using the CR method and the proposed PCCR method. The OTS results for CR and PCCR are shown in Table 7.13.

TABLE 7.10 Candidates with PCCR

Transmission Lines	PCCR ($/h)	CR ($/h)
BURR_OAK_34513_A_LN	35,722.98	305,032.27
DUMONT_DUMONSTILL34_1_1_LN	173.68	19,994.85
M_TOWN_M_TOWBLRST11_1_1_LN	**102.89**	**13,215.91**
FRANCESC_FRANCHANNA34_1_1_LN	**35.65**	**18,346.87**
BUNGE_BUNGEHASTI16_1_1_LN	35.30	25,242.07
EWINAMAC_13882_A_LN	21.09	42,393.93
EWINAMAC_13832_A_LN	15.68	95,546.81
CLARNDA_CLARNHASTI16_1_1_LN	**14.80**	**18,626.53**
EAU_CLA_EAU_CKING34_1_1_LN	9.26	109,724.77
ARGENTA_ARGENPALIS34_1_1_LN	**2.71**	**24,366.57**
ARGENTA_ARGENPALIS34_2_1_LN	**2.71**	**24,366.57**
AJ_MA_MO_AJ_MAMONRO34_1_1_LN	**2.17**	**58,639.49**
ARGENTA_ARGENTWIN_34_1_1_LN	1.71	70,111.24
MITCHLCO_MITCHHAZLT34_1_1_LN	**1.30**	**14,351.31**

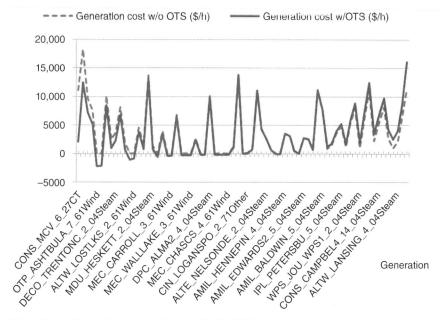

Figure 7.6   Generation costs without and with OTS.

TABLE 7.11   Comparison of objectives with and without OTS for CR and PCCR

	W/o OTS ($/h)	W/ OTS ($/h)	
		CR	PCCR
Objective ($/h)	888,471	876,997	864,197

TABLE 7.12   Comparison of objectives with and without OTS for CR and PCCR

	W/o OTS	W/ OTS	
		CR	PCCR
Open lines		2	5
Objective ($/h)	776,759	772,239	771,687

TABLE 7.13   Comparison of objectives with and without OTS for CR and PCCR

	W/o OTS	W/ OTS	
		CR	PCCR
Open lines		7	6
Objective ($/h)	1,286,753	1,151,641	1,142,653

As a result of OTS, seven lines are open using the CR method and six lines are open using the PCCR method. The objective saving of CR is about \$135,112/h equal to about 10.5% of saving; the objective saving of PCCR is about \$144,100/h equal to about 11.2% of saving. The PCCR has \$8988/h more saving than that of CR.

It is interesting to note that the saving of PCCR is more apparent than that of CR when the system is more stressed.

7.7 FINAL REMARKS

OTS can leverage grid controllability for enhancing system performance. Due to the scale of the problem, a limited set of transmission switching candidates is preselected to be potentially switched off by an energy and topology co-optimization algorithm for all practical purposes. A couple of criteria are presented and a heuristic method is proposed to select candidates for transmission-line switching in order to achieve a better market surplus within a reasonable market time frame. Simulation results from a large network have demonstrated that ranking transmission lines based on both CR and PC could be an effective way to select transmission switching candidates and gain objective savings. However, the proposed method does require the calculation of the sensitivity of bus angle with respect to the line flow prior to solving OTS. Further studies and evaluation of the impact of reserve costs on the selection of transmission line candidates are desirable in energy and ancillary services co-optimization markets.

As a final remark, besides computational performance concerns, it is important to note that OTS may likely affect locational energy prices and revenue adequacy in the financial transmission right (FTR) market when FTR settlements are financed by congestion revenues. FTR is a financial instrument to hedge against the uncertainty of transmission congestion charges. The FTR market could be highly vulnerable to manipulation of the grid topology. Although there are many benefits from OTS, it might undermine one of the fundamental assumptions upon which the FTR markets are based.

APPENDIX 7.A NOMENCLATURE

g	Index for generator
t	Index for study period
k	Index for transmission line
m,n	Index for "*from*," "*to*" bus of transmission line k
$\text{line}_m^{\text{to}}$	Set of transmission lines whose "*to*" side is connected to bus m
$\text{line}_m^{\text{fr}}$	Set of transmission lines whose "*from*" side is connected to bus m
l_t, l_{mt}	Load forecast for system and bus m for study period t
$\underline{p}_{gt}$	Low operating limits of generator g for study period t

$\overline{p}_{gt}$	High operating limits of generator g for study period t
$\underline{r}_t$	Reserve requirement for study period t
$\overline{r}_{gt}$	Reserve limit of generator g for study period t
B	Susceptance of transmission line
$\overline{f}_{kt}$	Limit of transmission line k for study period t
a_{kgt}	Sensitivity of transmission line k with respect to generator g for study period t
$\overline{\theta},\underline{\theta}$	High and low limits of bus angle for study period t
u_{gt}	Commitment status for generator g for study period t
p_{gt}	Power output of generator g for study period t
r_{gt}	Reserve contribution of generator g for study period t
f_{kt}	Transmission line flow for study period t
$\mathbf{u}$	Vector of all u_{gt}
$\mathbf{p}$	Vector of all p_{gt}
$\mathbf{R}$	Vector of all r_{gt}
Γ	Set of feasible solutions
$\chi_{gt}(\cdot)$	Production and no-load costs for generator g for study period t
$\varsigma_{gt}(\cdot)$	Start-up costs for generator g for study period t
$p_t^{\mathrm{loss}}(\cdot)$	System loss for study period t
$p_{mt}^{\mathrm{loss}}(\cdot)$	Loss at bus m for study period t

REFERENCES

[1] K. W. Cheung, "Ancillary Service Market Design and Implementation in North America: From Theory to Practice," Panel paper, Proceedings of the 3rd International Conference on Electric Utility Deregulation, Restructuring and Power Technologies (DRPT 2008).

[2] A. J. Wood and B. F. Wollenberg, *Power Generation, Operation, and Control*, 2nd edition, John Wiley & Sons, Inc., New York, 1996.

[3] J. H. Chow, R. deMello, and K. W. Cheung, "Electricity Market Design: An Integrated Approach to Reliability Assurance," Invited paper, *IEEE Proceedings* (Special Issue on Power Technology & Policy: Forty Years after the 1965 Blackout), vol. 93, pp.1956–1969, November 2005.

[4] K. W. Cheung, X. Wang, B.-C. Chiu, Y. Xiao, and R. Rios-Zalapa, "Generation Dispatch in a Smart Grid Environment," Presented at the 2010 IEEE/PES Innovative Smart Grid Technologies Conference (ISGT 2010).

[5] H. Li, G. W. Rosenwald, J. Jung, and C.-C. Liu, "Strategic power infrastructure defense," *Proceedings of the IEEE*, vol. 93, no. 5, pp. 918–933, 2005.

[6] E. B. Fisher, R. P. O'Neill, and M. C. Ferris, "Optimal transmission switching," *IEEE Transactions on Power Systems*, vol. 23, no. 3,, pp. 1343–1355, August 2008.

[7] K. W. Hedman, R. P. O'Neill, E. B. Fisher, and S. S. Oren, "Optimal transmission switching—sensitivity analysis and extensions," *IEEE Transactions on Power Systems*, vol. 23, no. 3, pp. 1469–1479, August 2008.

[8] J. D. Fuller, R. Ramasra, and A. Cha, "Fast heuristics for transmission line switching," *IEEE Transactions on Power Systems* , vol. 27, no. 3, pp. 1377–1386, August 2012.

[9] P. A. Ruiz, J. M. Foster, A. Rudkevich, and M. C. Caramanis, "On Fast Transmission Topology Control Heuristics," Proceedings of 2011 IEEE PES General Meeting.

[10] A. Khodaei, M. Shahidehpour, and S. Kamalinia, "Transmission switching in expansion planning," *IEEE Transactions on Power Systems*, vol. 25, no. 3, pp. 1722–1733, August 2010.

[11] A. Khodaei and M. Shahidehpour, "Security-constrained transmission switching with voltage constraints," *International Journal of Electrical Power and Energy Systems*, vol. 35, no. 1, pp. 74–82, February 2012

[12] A. Khodaei, M. Shahidehpour, and Y. Fu, "Transmission switching in security-constrained unit commitment," *IEEE Transactions on Power Systems*, vol. 25, no. 4, pp. 1937–1945, November 2010.

[13] K. W. Cheung, G. W. Rosenwald, X. Wang, and D. I. Sun, "Restructured electric power systems and electricity markets," in *Restructured Electric Power Systems: Analysis of Electricity Markets with Equilibrium Models*, X.-P. Zhang, Ed., Wiley-IEEE Press, 2009, pp. 53–98.

[14] K. W. Cheung, J. Wu, and R. Rios-Zalapa, "A Practical Implementation of Optimal Transmission Switching," Presented at the 4th International Conference on Electric Utility Deregulation, Restructuring and Power Technologies (DRPT 2011).

[15] K. W. Cheung, "Economic Evaluation of Transmission Outages and Switching for Market and System Operations," Panel paper, Proceedings of 2011 IEEE PES General Meeting.

[16] J. Wu and K. W. Cheung, "On Selection of Transmission Line Candidates for Optimal Transmission Switching in Large Power Networks," Proceedings of 2013 IEEE PES General Meeting.

CHAPTER 8

TOWARD VALUING FLEXIBILITY IN TRANSMISSION PLANNING

Chin Yen Tee and Marija D. Ilić

THE INCREASED PENETRATION of distributed and renewable energy sources is expected to increase the level of uncertainty and variability in the grid, which would bring about a greater need for flexibility in the system. In order to provide more flexibility in the system, both the transmission infrastructure and regulatory framework need to be enhanced to provide transmission owners and operators with the flexibility required to maintain the reliability and efficiency of the grid. In recent years, various technologies have been developed to allow for greater flexibility in the transmission system. However, the current transmission operational, planning, and regulatory frameworks are not designed to adequately account for the value of flexibility. In this chapter we explore how the disconnect among the current operational, planning, and market mechanisms hide the value and potential of flexible technologies. In addition, we illustrate how flexible technologies can change the scale economies of the transmission sector, with the support of a well-designed planning framework. Such planning framework should provide mechanisms to account for the value of operational flexibility in transmission planning and to provide market mechanism for information and risk sharing among stakeholders. Finally, we highlight some key challenges toward developing transmission rights for transmission flexibility.

8.1 INTRODUCTION

Environmental and climate concerns, technological developments, and regulatory changes have changed the electricity industry landscape in the United States in recent years. On the generation side, environmental and climate concerns, along with regulatory changes, have resulted in an increased penetration of renewable energy sources. On the demand side, the introduction of new technologies, such as smart metering devices and smart home appliances, empowers customers with greater control over their electricity usage. Such changes in the industry landscape are expected to

Power Grid Operation in a Market Environment: Economic Efficiency and Risk Mitigation, First Edition.
Edited by Hong Chen.
© 2017 by The Institute of Electrical and Electronics Engineers, Inc. Published 2017 by John Wiley & Sons, Inc.

continue in the near future due to continuing concerns over climate change and technological development that improve the cost effectiveness of new technologies.

One key impact of the changing electricity mix is the increased level of uncertainty and variability seen by the grid. In the short run, the intermittent and variable renewable energy resources increase short-run uncertainties in the grid. Similarly, long-run uncertainties that are likely to be brought about by deployment of new disruptive technologies as well as by unexpected changes in generation expansion plans make it hard to invest in transmission at value. This creates complex challenges to both system operators and planners responsible for reliable and efficient power delivery. It is particularly difficult to plan and operate novel, flexible technologies that could potentially allow for more flexible and efficient operation of the system. In order to operate and plan the electricity grid in a reliable and efficient manner, there is a need to develop a systematic methodology to not only appropriately characterize these uncertainties, but also integrate considerations of these uncertainties in operational and investment decision making in the electricity industry.

Traditionally, flexibility has not been a major option in the operation and design of the transmission infrastructure. The current transmission investment planning framework focuses on building of new transmission lines to meet the need of changes in energy demand and does not adequately consider alternatives to building additional transmission line capacity. This lack of consideration for alternative technologies is unfortunate as in recent years, development in fast power electronics switches has led to the availability of a large variety of control, communication, and sensing technologies that could improve flexibility in the transmission sector and potentially reduce the need for new transmission lines. As it becomes more difficult and costly to build conventional transmission lines due to right-of-way issues and public resistance, the transmission sector should be encouraged and incentivized to explore alternative technologies that could potentially ease the need for new transmission line capacity.

However, the existence of flexible technologies alone is insufficient to ensure flexibility in transmission operation and planning. Transmission flexibility needs to be supported by operational, regulatory, and market framework that appropriately values flexibility in the grid and that supports rather than hinders the use of flexible technologies. The conventional operational and planning framework for the electricity industry was designed with conventional technologies in mind and generally does not consider previously unproven disruptive technologies such as flexible line flow control. Under the current operational framework and regulatory rules, the value of flexibility is often hidden and is not explicitly accounted for within the given institutional design. Therefore, a serious rethinking of regulatory, policy, and institutional design in the electricity industry is needed to provide appropriate incentives for flexibility in the transmission sector.

The goal of this chapter is to highlight impediments toward flexibility in the transmission sector and to recommend changes to current electricity policy and institutional framework to provide better incentives for flexibility. We begin by considering how flexible technologies could potentially change the scale economies of the transmission sector. Next, we evaluate the current disconnect among the operational, planning, and market frameworks in the electricity sector that impedes the integration of flexible technologies. We illustrate how different operational frameworks can have different impacts on investment decisions and highlight the importance of considering the operational and market effects when making investment decisions. We also highlight the need for well-designed long-term markets to encourage information and risk sharing among stakeholders. Finally, we relate key challenges of developing transmission rights for supporting long-term market incentives to invest in transmission.

8.2 SCALE ECONOMIES OF TRANSMISSION TECHNOLOGIES

The well-recognized argument for the heavy regulation of the transmission sector is that the electric transmission grid represents a natural monopoly due to the economies of scale and lumpiness of transmission infrastructure. However, similar to how technological developments have reduced economies of scale in the generation sector, the introduction of new, flexible technologies in the transmission sector has the potential of changing the scale economies of the transmission sector. For example, line power flow control devices, such as phase angle regulators (PAR) and thyristor-controlled series capacitors (TCSC), can be used to adjust the electrical parameters of the system, especially when there exists significant unused transmission line capacity in the system. These technologies allow for incremental investments in line flow control technologies to be made. To illustrate this, we consider a technology that controls line power flow by changing the line reactance. Such flexible reactance can be modeled as a controllable capacitor placed in series with the transmission line (Figure 8.1). The controllable capacitor changes the admittance matrix of the transmission network, which, in turn, alters the pattern of line power flow in the system. For a single line with flexible reactance, the line impedance can be written as

$$Z_{\text{line}} = R_{\text{line}} + j(X_{\text{line}} - X_{\text{Flex}})$$ (8.1a)

Changes in the reactance of a line in the system affect the line power flow distribution in the system according to Kirchhoff's current law (KCL) and Kirchhoff's voltage law (KVL). The effects of variable reactance on line flows in the grid can be illustrated using a small two-node electric power system comprising two parallel lines whose reactances are X_A and X_B, respectively. It can be shown that under the "DC" power flow formulation, KCL causes real power flow in the two lines to split in inverse proportion to the proportion of reactance in each branch [1]. Mathematically,

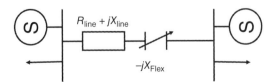

Figure 8.1   Model of line with device providing flexible reactance.

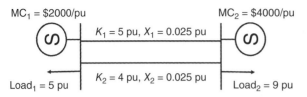

Figure 8.2   Base case system for example to show investment effects of flexible reactance.

the real power line flows (F_A and F_B) through two parallel lines carrying a combined power of P are

$$F_A = \frac{X_B}{X_A + X_B} P \tag{8.1b}$$

$$F_B = \frac{X_A}{X_A + X_B} P \tag{8.1c}$$

The potential of flexible reactance devices in providing more flexibility in making transmission investment decision can be illustrated via a simple example. In the two-bus base case system shown in Figure 8.2, two nodes are connected via two transmission lines with equal reactance but different thermal capacity limits. The cheaper generation is located on the node on the left. To simplify the exposition, we assume that the generator limits are not binding constraints in this system.

In such a system, the most economically efficient power dispatch is to try to transfer as much power as possible from the cheaper node to the more expensive node. In the base case, the maximum power that can be transferred from the cheaper node to the more expensive node is 8 pu if we consider only the thermal capacity limit. Even though line 1 has a thermal limit of 5 pu, the maximum power that can flow through line 1 is still 4 pu (i.e., the thermal limit of line 2) due to KCL.

Now, consider the following question: What do we need to do to increase the power transferred from the cheaper node to the more expensive node to 9 pu? First, we consider potential investment in new transmission line (Figure 8.3).

Assuming that the reactance of the new line will be the same as the reactance of the original lines, the minimum new line investment of 3 pu is needed to increase the transfer capacity to 9 pu. With a new line investment of 3 pu, the maximum power flow in each of the three lines will be 3 pu. Given that transmission line investment is lumpy, it is likely that a line whose thermal capacity is higher than 3 pu will have

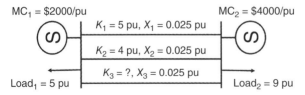

Figure 8.3   Transmission line capacity needed to increase transfer capacity to 9 pu.

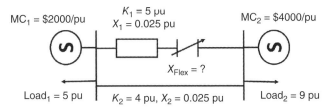

Figure 8.4   Flexible reactance investment needed to increase transfer capacity to 9 pu.

to be installed to provide the increase in transmission capacity. The lumpy nature of transmission investment and the economies of scale of such investment are often cited as reasons why merchant transmission investments are inefficient [2].

Now, in Figure 8.4, we consider potential investment in devices that provide flexible reactance to reach the same goal. In this case, such devices can be used to lower the reactance of line 1 to enable more of its line capacity to be used:

In this case, a device that provides 0.005 pu of flexible reactance will lower the reactance of line 1 to 0.02 pu. Using equations (8.2) and (8.3), this will increase the maximum flow in line 1 to its thermal limit of 5 pu, while the maximum flow in line 2 remains at 4 pu.

The key difference between investment in flexible reactance device and investment in new transmission line is that investment in flexible reactance device is not lumpy, which allows us to make marginal expansion in transmission capacity. In addition, investment in flexible reactance devices allows us to avoid right-of-way cost that often plagues transmission line deployment. To further demonstrate the changing scale economies of transmission investment, we estimate the cost of the two investment options presented in the example above.

The cost of new transmission investment is dependent on the voltage rating, the number of circuits, and the length of the transmission line. The limiting factor for the operational capacity of the transmission line is dependent on the voltage rating and the length of the line. Lines with higher voltage ratings tend to have higher capacity. Shorter lines (<50 miles) tend to be limited by the thermal capacity limit of the line, whereas longer lines (>50 miles) tend to be limited by the surge impedance loading limits [3]. Figure 8.5 shows some typical cost for new transmission line investments using data taken from [3,4], which include the cost of obtaining right of way. Information about the cost of FACTS devices is difficult to obtain. For devices that provide flexible reactance, a cost of $135,000/MVar has been cited in literature [5]. The MVar operating range of the flexible reactance is given by the following equation [5]:

$$s = X_{\mathrm{c}} \frac{K_{\mathrm{line}}^2}{S_{\mathrm{base}}} \tag{8.2}$$

where X_{c} is the maximum series capacitor reactance in pu, K_{line} is the thermal line limit of the transmission line, and S_{base} is the MVA base power (100 MVA in the examples shown in this chapter).

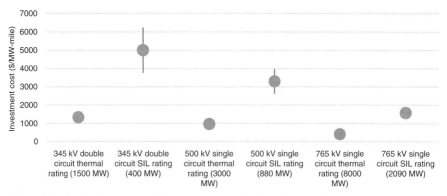

Figure 8.5   Typical cost ranges for new transmission line investments.

The cost of investments of the two options presented in the example above is estimated for different assumptions of length and voltage rating of transmission lines. In the graph in Figure 8.6, the cost of building the 300 MW new line in the example is presented for lines of different lengths, using the lowest $/MW-mile line investment cost shown in Figure 8.5. The cost of installing the flexible reactance device in line 1 is also calculated based on per-mile line reactance for 345 and 765 kV transmission lines given in [6]. Based on the graph, it can be seen that for transmission line of all lengths, the flexible reactance is the lower cost option for achieving the same increase in transmission capacity. If we assume the higher $/MW-mile line investment cost shown in Figure 8.5, the results would favor the flexible reactance device even more.

The relative cost of investing in new transmission lines and investing in flexible devices to achieve the same increase in capacity is dependent on the current configuration in the system. Flexible devices are useful in systems with low utilization of existing lines. New transmission lines are useful when larger capacity expansion is needed. A comprehensive cost analysis of the different investment options needs to be done on a case-by-case basis. In closing, new technologies have changed the

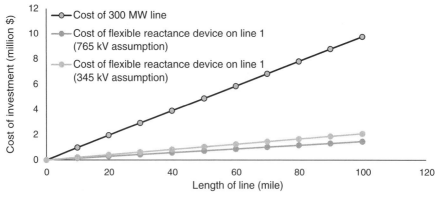

Figure 8.6   Comparison of cost of new line investment and cost of investment in flexible reactance device.

economics of transmission investment. Changes in scale economies in the generation sector has allowed for greater competition and market-based solutions in the generation sector, suggesting that market-based solutions could play a greater role in the transmission sector, especially if the appropriate institutional framework exists to support such solutions.

8.3 DISCONNECT OF CURRENT POWER SYSTEM OPERATIONAL, PLANNING, AND MARKET MECHANISMS

Current institutional framework supporting the electricity industry generally does not support flexibility in the transmission sector. Several different reasons for this are described next. The first reason is the decoupling between operations and planning. Current infrastructure planning is often done for worst case scenarios, without considerations of how real-time operational flexibility can be used to manage these worst case scenarios during actual system operation. Large infrastructure investments are often made to handle scenarios that rarely occur, resulting in infrastructures that are underused most of the time. This is likely to worsen with increasing level of renewable energy penetration as the transmission lines used to support renewables are typically sized based on the maximum generating capacity of the renewable [7], even though the capacity factors for intermittent energy sources such as wind and solar energy are typically only 20–40% of the rated capacity [8]. Such large transmission infrastructure investment could be avoided if we make transmission investment and sizing decision at value and consider potential options for operational flexibility, such as demand response and storage technologies, in the system. In order to do so, the operational framework of the system needs to explicitly be taken into account in transmission planning.

Second, current electricity industry also suffers from a disconnect between electricity markets and electricity planning and operation. Mismatch between the constraints considered in electricity markets and the actual physical constraints during real-time operations has resulted in a need for significant uplift payments in the energy market [9].

Third, in markets for financial transmission rights (FTR), mismatch between FTR capacity and real-time operational transmission capacity has resulted in revenue inadequacy in the system [10]. These inconsistencies between market prices and actual values make it difficult for stakeholders to effectively manage risk in both the short and long runs. In addition, it provides inaccurate and inefficient signals for investments in both conventional and flexible technologies.

8.4 IMPACT OF OPERATIONAL AND MARKET PRACTICES ON INVESTMENT PLANNING

The decoupling among operation, market, and planning is problematic as the value of different technologies depends on the supporting operational and market framework.

In particular, rigid operational and market mechanisms, such as the use of preventive $(N-1)$ security-constrained dispatch and the treatment of load as inelastic, favor conventional technologies and hide the value of flexible technologies. In contrast, flexible operational and market mechanisms that enable operational flexibility incentivize flexible technologies.

A simple two-bus example is used to illustrate the impact of different operational and market practices on investment decision making. The four different operational and market approaches considered here are as follows:

1. Economic dispatch, where the only goal is to minimize generation dispatch cost. Reliability needs are not considered in this case.

2. Preventive $(N-1)$ security-constrained economic dispatch, where power generation is dispatched such that the system will remain secure even if any single system element fails. This dispatch method is commonly used in current system operation. The key drawback of this dispatch framework is that it leads to highly conservative and economically inefficient dispatch. In this chapter, we consider potential failures in transmission line elements only.

3. Corrective $N-1$ security-constrained economic dispatch, where the potential for corrective control actions and re-dispatch is accounted for. This is the operational framework that is frequently touted by smart grid proponents as it accounts for the value of corrective control technologies. The main arguments in support of corrective $N-1$ security-constrained economic dispatch is that the system can often remain operational for a short period of time after outages occur, which allows for fast corrective actions to be taken to bring the system within the acceptable operational limits [11].

4. Corrective $N-1$ security-constrained economic dispatch with elastic load. This is similar to the previous case except that load is considered to be dispatchable in this case. Each load is assigned a "loss-of-load" price, which represents the amount of compensation that the load is willing to receive in exchange for load being shed during contingencies.

8.4.1 Case Study A

In this case study, there is an urban location and a remote location connected via two transmission lines (Figure 8.7). The urban location has a higher maximum load of 2000 MW or 20 pu, while the remote location has a lower maximum load of 500 MW or 5 pu. In addition, the urban location has two types of gas turbines (a 2500 MW, $30/MWh gas turbine and a 2500 MW, $50/MWh gas turbine), while the remote location has a coal power plant (500 MW, $20/MWh).

Now, assume that a new environmental regulation is being considered that will cause the coal plant to be shut down in the near future. In anticipation of the coal plant being closed, a wind power developer is considering building a new 2000 MW wind farm at the remote location. In addition, a tech company is considering building

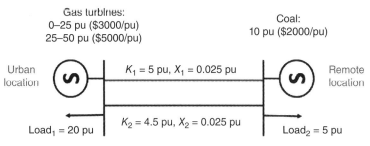

Figure 8.7   Base case system for case study.

a large data center at the remote location, which would increase the load at the remote location to 1000 MW. This is illustrated in Figure 8.8.

The key question to be answered here is: What is the optimal level of transmission line and flexible reactance device investment in anticipation of these changes under different operational and market frameworks? The modeling approach and mathematical model used to evaluate this question is shown in Appendix 8.B and the investment cost used in this chapter is shown in Appendix 8.C. The results of the implementation of the modeling approach are presented next.

The first step to valuing investment decision is to characterize the uncertainties being considered. A sound understanding of the nature of the uncertainty and the information available to characterize the uncertainties is needed to select an appropriate method to characterize uncertainties. In this case study, short-run wind, load fluctuations, and transmission element outages are considered. Each transmission element is assumed to have a 2% chance of failure. Historical hourly wind and load data from PJM for the years 2012 and 2013 were used to obtain wind and load fluctuation patterns for this case study [12]. The urban location uses load data from the PJM Mid-Atlantic region, while the remote location uses wind and load data from the PJM West region. The K-means clustering technique was used to produce 100 scenarios that together capture the wind and load fluctuations and correlation among the data. K-means clustering technique is used as it allows us to capture the variation of wind and load levels at each node and also the correlation among wind and load levels at different nodes. This clustering technique is described in detail in Appendix 8.B.

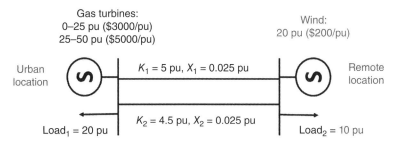

Figure 8.8   System with anticipated changes.

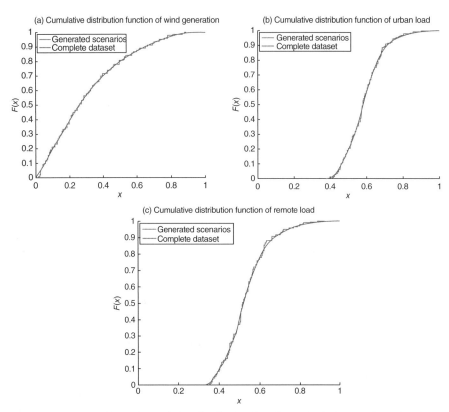

Figure 8.9 Comparison of cumulative distribution function: (a) wind generation, (b) urban load, and (c) remote load.

In order to evaluate the performance of the K-means clustering algorithm in providing a reduced model of the dataset, the cumulative distribution function and correlation matrix of the 100 generated scenarios were compared to that of the complete dataset. The comparison of the cumulative distribution function for the wind generation, urban load, and remote load is shown in Figure 8.9.

The Spearman correlation coefficient is used to compare the rank correlation among the generated scenarios and the complete dataset. The correlation matrix for the complete dataset and that for the generated scenarios are shown in Table 8.1. The

TABLE 8.1 Comparison of correlation matrix of complete dataset and generated scenarios (generated scenarios correlation in parenthesis)

	Wind	Urban Load	Remote Load
Wind	1 (1)	–0.09 (–0.09)	–0.16 (–0.16)
Urban load	–0.09 (–0.09)	1 (1)	0.92 (0.96)
Remote load	–0.16 (–0.16)	0.92 (0.96)	1 (1)

TABLE 8.2   Results of simulation for optimal investment considering short-run operational flexibility

Operational Framework	Investment Decision	Operational and Annualized Investment Cost
Economic dispatch	FRD line 1 (0.0025 pu)	$0.303 billion
Preventive $N-1$	Line (450 MW)	$0.314 billion
Corrective $N-1$	Line (353 MW), FRD both lines (0.0053 pu)	$0.311 billion
Corrective $N-1$ with dispatchable load	FRD line 1 (0.0025 pu)	$0.307 billion

FRD, flexible reactance device.

comparison of both the cumulative distribution functions and correlation matrices confirms that the set of scenarios generated via K-means clustering is a good representation of the original data.

Having generated the representative scenarios using K-means clustering, we implemented model 1 in Appendix 8.B for the four different operational approaches outlined earlier. The results of the simulations are summarized in Table 8.2.

In the first case, we do not consider the reliability needs of the system. Under the optimization approach, investments are only made if the cumulative value of the generation cost savings is greater than the cost of investment. The optimal investment option is to invest in 0.0025 pu of flexible reactance in line 1, which allows for the full 500 MW capacity of line 1 to be used. This entails a total operational and investment cost of $0.303 billion/year.

In the second case, we consider the preventive $N-1$ security-constrained economic dispatch. In this case, the optimal investment decision is to invest in a new 450 MW transmission line. This line is needed to maintain enough transmission capacity when any of the other two lines fail. As expected, this results in the most conservative and expensive economic option.

In the third case, we account for the ability of the system to take corrective actions to maintain the reliability of the system during contingencies. In this case, a transmission line with a smaller capacity is required. Flexible reactance devices are installed in both the existing lines to ensure that the full capacity of those lines can be used when any of the other lines fails. This scenario cost slightly less than the second scenario.

Finally, in the fourth case, we account for the ability to shed load during contingencies. The "loss-of-load" price assigned to the urban load is $5000/MWh whereas the price assigned to the remote load is $1000/MWh. Due to the additional flexibility brought about by the ability to dispatch loads during contingencies, the optimal investment decision is identical to the investment decision for the purely economic case (i.e., invest in flexible reactance in line 1).

The results demonstrate the importance of accounting for the underlying operational and market framework when making investment decisions. In particular, the results demonstrate that rigid operational and market frameworks, such as preventive

$N - 1$ security-constrained economic dispatch and the treatment of load as inelastic, result in inefficient line overinvestment. As it stands, the value of many flexible technologies are hidden and lost under the current conservative and rigid operational framework, which deters investors from investing in such technologies. In order to provide appropriate incentives for investment in flexible technologies and transition toward a smart grid, it is critical that we not only adopt a more flexible operational and market framework, but also account for operational and market flexibility during investment planning.

8.5 INFORMATION AND RISK SHARING IN THE FACE OF UNCERTAINTIES

Even if we account for operational and market flexibility in making investment planning, the ability to make efficient investment decision in the transmission sector is still limited due to the difficulty of making accurate long-term predictions in the electricity sector. Long-term load predictions done by system operators and transmission planners have been demonstrated to be highly inaccurate [13]. In addition, generation expansion plans have been known to change suddenly and unexpectedly. Historical data are insufficient in making accurate long-term predictions due to the rapidly changing technological landscape in the electricity industry. This uncertainty in generation and load patterns can result in inefficient overinvestment or underinvestment. Under a guaranteed rate-of-return regulatory framework, consumers bear the risk of inefficient investments and there are no true incentives or mechanisms for transmission planners to improve the way they deal with uncertainty in transmission planning.

We extend the case study presented earlier to illustrate the impact of long-term uncertainty on investment decisions.

8.5.1 Case Study B

Now, we extend the previous case study and consider the fact that policy changes and business plans are often uncertain. In this case, we assume that there is a possibility that the environmental policy did not get approved by Congress. If it does not get approved by Congress, the coal plant remains open and the new wind farm is not built. In addition, the tech company is not sure how much energy will be needed by the new data center. The two uncertain state variables here are the generation capacity at the remote region (GC) and the maximum load level at the remote region (ML). The potential states for these uncertain state variables are

$$GC \in \left\{ \begin{array}{l} \text{Coal Remains, No Wind (GC1/Base)} \\ \text{2000 MW Wind Generation (GC2)} \end{array} \right\}$$

$$ML \in \left\{ \begin{array}{l} \text{No Increase in Load (ML1/Base)} \\ \text{Load Level Increase to 1000 MW (ML2)} \\ \text{Load Level Increase to 1500 MW (ML3)} \end{array} \right\}$$

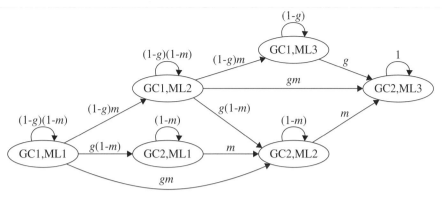

Figure 8.10   State transition matrix for uncertain states.

The state transition matrix for the uncertain states are shown in Figure 8.10. In this figure, m represents the probability that the load level will increase from ML1→ML2 and from ML2→ML3, while g represents the probability that the wind farm will be built. Notice that the model could support different transitional probabilities from ML1 to ML2 and from ML2 to ML3.

A Markov decision process (MDP) is used to solve this long-term dynamic investment problem. The details of the mathematical formulation can be found in model 2 of Appendix 8.B. A 10% per annum discount rate is used. The operational cost at each state was calculated using the same 100 wind and load scenarios found through K-means clustering. The wind and load levels for the different scenarios are scaled to correspond to the appropriate states. In the cases with inelastic load, the system is penalized heavily for not meeting the load, such that the system will always seek to meet the load. The investment decisions being considered are investment in 0.0025 pu of flexible reactance in line 1 and a new 450 MW new transmission line. The results of three different cases are described next.

First, we consider the deterministic case where $g = 1$ and $m = 1$. The operational approach considered here is the economic dispatch only approach. In this case, we know for sure that the demand will increase from 500 to 1000 MW at the remote node between decision periods 1 (year 1) and 2 (year 5), and from 1000 to 1500 MW between decision periods 2 and 3 (year 9). In addition, the wind farm will be built between decision periods 1 and 2. In this case, the optimal investment decision is to invest in 0.0025 pu of flexible reactance in line 1 during year 1 and invest in the 450 MW new line in year 9.

For the second case, we consider a case similar to the first case. However, in this case $g = 0.5$ and $m = 0.5$. In this case, future evolution of demand and generation investment is uncertain. The optimal investment decision in year 1 here is to build the 450 MW new line. The investment decision in years 5 and 9 will be dependent on the actual realization of the demand and generation investment. The flexible reactance device will be built in year 5 or year 9 only if the demand increased from the base case. In this case, it can be observed that uncertainty in future demand growth and generation investment pattern resulted in more aggressive and expensive investment decision. One key reason for this conservative investment decision is due to the assumption that demand is inelastic.

In the final case, we use the same assumptions as the second case ($g = 0.5$, $m = 0.5$). However, load is considered to be elastic in this case. Similar to the earlier simulations, we assume that the "loss-of-load" price for the urban load is $5000/MWh whereas the "loss-of-load" price for the remote load is $1000/MWh. In this case, the optimal investment decision in year 1 is to invest in the flexible reactance device. The new line will only be built in year 5 or year 9 if the load increased to 1500 MW and the wind farm is built. This demonstrates how accounting for load elasticity gives us the flexibility to invest in smaller, flexible devices as an intermediate measure in face of long-term uncertainty. This enables us to delay the decision to invest in the more expensive transmission line until more information on future load and generation pattern can be obtained.

The results above highlight two key requirements for efficient transmission investment:

1. Load needs to participate in the market, both in the long run and in the short run. In current long-term transmission planning, transmission investment is made to fully accommodate all future load growth without considering how much customers are willing to pay for the delivery of the energy. This is despite the fact that loads are more elastic in the long run as customers can take steps to reduce their load consumptions through energy efficiency programs or installation of private generators. The results demonstrate that if we demand that total load has to be met at all cost, transmission planner will tend to overinvest to account for the worst case load scenario. In addition, the potential of incremental expansion in transmission capacity using smaller, flexible devices is often ignored when the incentive is to overdesign the system.

   Accounting for load elasticity allows more flexibility to transmission planners, by enabling them to make incremental, smaller transmission upgrades/investments while waiting for more information on future load and generation pattern to arrive. Technologies are available to allow for more fine grained dispatched of demand. However, in order to enable greater participation of load in electricity markets, some sort of market mechanism needs to exist for load to make their long-term willingness-to-pay function for electricity or willingness-to-accept-compensation function for load shedding be known. One way in which this can be done is to design electricity markets that are symmetric (i.e., load submits bids along with generation to reveal how much they are willing to pay for electricity). Another potential way in which this can be done is to put in place a menu of reliability insurance program where load can purchase different level of reliability and be compensated at different price level, should load be shed. Regardless of how it is done, there needs to be a way to allow load to participate and express its preference in electricity markets of all timescales.

2. There needs to be a mechanism for long-term information to be exchanged among stakeholders. Long-term uncertainty has a significant impact on investment decisions. The key question to be answered here is: How can we empower transmission owners and planners with more accurate information about future load and generation uncertainties and allow for long-term risks to be shared among stakeholders?

One potential strategy that we propose is to design long-term markets for transmission (e.g., 5 years ahead) and to require compulsory participation in the market by load-serving entities and generation. Load-serving entities should be required to procure their long-term forecasted electricity demand that should be within a certain range of the actual realized demand. For example, load-serving entities could be required to procure sufficient generation 5 years ahead such that the actual required demand falls within plus or minus 15% of the procured demand. In order to encourage load-serving entities to provide as accurate a forecast as possible, the load-serving entities would be penalized if the demand falls outside the plus or minus 15% range. The rationale behind this penalty is that the load-serving entities should be penalized for the underinvestment or overinvestment that occurs due to vastly inaccurate forecasts. A similar regulation should be put in place for generation, where generation is required to contract out a certain percentage of its generation via long-term contracts and be penalized for not meeting those needs. A preliminary framework showing the information exchange for the long-term market is shown in Figure 8.11, where λ_{forward} is the forward price of electricity and π is the future marginal penalty that the generation or load-serving entities will face if the contracted quantity in the forward market is vastly different from the spot. For example, the penalty to be paid by

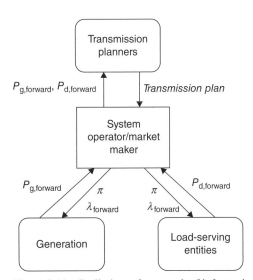

Figure 8.11   Preliminary framework of information exchange for long-term market.

the load-serving entity when real-time market has cleared is given by the following equation:

$$\pi \, \max(|P_{d,forward} - P_{d,spot}| - \epsilon, 0) \tag{8.3}$$

where $P_{d,forward}$ is the contracted long-term demand, $P_{d,spot}$ is the actual realized demand during the spot market, and ϵ is the forecast error tolerance (e.g., 15% of $P_{d,forward}$). The load-serving entity will only be penalized if the forecast is off by a certain tolerance level.

The benefit of such a system is that it provides bounds to the uncertainty that transmission planners face in making investment decision. Information regarding the bounds on uncertainty and on how much customers are willing to pay for demand can allow transmission planner to make more efficient and flexible investment decision. In addition, the cost of underinvestment or overinvestment is shared by the stakeholders. The framework described above is a preliminary proposal and more work needs to be done to provide a more detailed evaluation of different market design for long-term markets in the electricity sector.

8.6 CHALLENGES IN DESIGNING FINANCIAL RIGHTS FOR FLEXIBILITY

In the preliminary framework shown in Figure 8.11, a centralized market maker is required to provide price signals to stakeholders and clear the market. The next open question and design challenge is how would we design FTRs such that they not only encourage efficient operation of the system, but also incentivize economically efficient investment in both flexible and conventional transmission technologies? Can we design a system of short-term and long-term transmission rights that provide appropriate price signals for investments and potentially allow for greater participation of merchant transmission investments?

The inefficiency of current FTRs implementation can be seen from the revenue shortfall that frequently occurs due to the mismatch between the system constraints considered during FTRs auction and real-time system condition. Currently, this revenue inadequacy is dealt with via "uplift" payments to owners of FTRs that are funded either through surplus revenue from periods of excess congestion rent or through transmission owners who pass the cost over to consumers via access charges [14]. In cases where the "uplift" payments are funded by surplus revenue, holders of FTRs are not guaranteed full payment. One main cause of this revenue inadequacy issue is the lack of consideration for both short-run and long-run uncertainties during FTRs auctions and allocations, which are commonly implemented via deterministic approaches [10]. Even without flexible technologies, the electricity transmission network is not static due to line de-ratings and outages. With the inclusion of flexible technologies that change the topology and electrical properties of the network, the revenue inadequacy problem brought about by the current FTRs implementation is likely to worsen. By neglecting the stochastic nature of the power system, FTRs ignore the value of flexibility and do not provide efficient signals for optimal investments in both flexible and conventional technologies. A key open requirement for the design of financial rights for flexibility is to develop market mechanisms to manage the uncertainties inherent in electricity operations [15]. At the very minimum, we believe that this would involve the creation of multi-timescale reconfiguration markets for FTRs that go as close to real time as possible, to enable market participants to adjust their transmission rights portfolio as new information is available on the anticipated system configuration.

The theoretical development of FTR has traditionally focused on its use as a hedging mechanism for congestion cost. In recent years, there has been increased interest in the use of long-term FTRs to encourage efficient investment decisions (e.g., [16, 17]). Henze et al. [17] present the results of several experiments on using long-term FTRs for inducing efficient investment decision and find that price signals from long-term FTRs do not provide consistent and efficient signals for investment and much is left up to the discretion of the transmission planner to interpret the price signal. Rosellon [16] combines concepts from long-term FTRs literature with concepts from performance-based regulation literature to propose a combined merchant-regulatory framework for transmission. The proposed mechanism has the potential of being a highly flexible institutional framework for transmission investment as, depending on the nature of the transmission investment, a merchant-based approach or performance-based regulatory approach might be more suitable. Despite some interesting results that have been presented in current literature, there remain many open research questions as to how long-term FTRs need to be designed to provide efficient price signals for investment. The design of well-functioning long-term financial rights mechanism is a second key challenge for the design of financial rights for flexibility.

8.7 CONCLUSIONS

With the increasing level of uncertainty in the electricity industry, power system operators and planners need to be equipped with the infrastructure and operational tools that will provide sufficient flexibility to continue to manage the electricity grid in an efficient and reliable manner. There are various control, communication, and sensing technologies that can provide more flexibility in the power system. However, the current market, policy, and regulatory framework in the electricity industry does not provide sufficient incentives for the efficient investment and use in such technologies. In fact, as demonstrated in this chapter, many of the current market, policy, and regulatory frameworks impede the introduction of new, flexible technologies.

In this chapter, we illustrate how flexible technologies have reduced the economies of scale of transmission sector. Models were developed to account for the value of operational flexibility and investment flexibility in making investment decisions. These models were used to demonstrate impediments to flexibility in the transmission sector and to make policy recommendation. We demonstrated the importance of not only adopting flexible operational and market framework, but also accounting for these flexibilities in making investment decisions. In addition, we demonstrated the need for load to participate more actively in electricity markets, both in short-term markets and long-term markets, and presented a preliminary framework for how a long-term market can be designed to provide more information for system planner and to enable long-term investment risk sharing. Finally, we identified some challenges in designing financial rights for transmission flexibility.

APPENDIX 8.A NOMENCLATURE

Z_{line}	Line impedance
R_{line}	Line resistance
X_{line}	Line reactance
X_{Flex}	Flexible reactance
s	MVar operating range of flexible reactance device
X_c	Maximum series capacitor reactance in pu for flexible reactance device
K_{line}	Thermal line limit of the transmission line in megawatts (new/existing line)
S_{base}	MVA base power
g	Transitional probability for wind level
m	Transitional probability for load level
$\lambda_{forward}$	Forward price of electricity
π	Future marginal penalty that the generation or load-serving entities will face if the contracted quantity in the forward market is vastly different from the spot
$P_{d,forward}$	Contracted long-term demand
$P_{d,spot}$	Actual realized demand in the spot market
$\in$	Forecast error tolerance
x	Operational decision variables
y	Investment decision variables
C_{opt}	Hourly operational cost
C_{invt}	Annualized investment cost
$Ct(s)$	Number of hours scenario s happens in a year
$N_s, N_G, N_{\bar{l}}, N_T, N_D$	Number of scenarios, number of generators, number of new lines, number of flexible devices, and number of dispatchable load
$c_g, c_{invt,\bar{l}}, c_{invt,t}, c_D$	Generation cost, new line investment cost, new flexible reactance cost, and price of loss load
P_g^s	Power generation of generator g for scenario s

x^s_{Flex}	Operational setting of the flexible reactance device in terms of change in reactance for scenario s
θ^s_n	Nodal power angle at node n for scenario s
f^s_{line}	Line flows for scenario s
K_{Flex}	New flexible reactance capacity
b_{line}	Binary variable indicating whether the new line is built
G	Binary matrix indicating the nodal location of the generators
S	Matrix that is -1 for a transmission line exiting a node and $+1$ for a transmission line entering a node
$P^s_g, f^s_{\text{line}}, D^s$	Vectors of power generation, transmission line flow, and nodal power demand for scenario s
$\theta^s_{l,\text{to}}, \theta^s_{l,\text{from}}$	Nodal power angles of the existing and entering node of the transmission line l for scenario s
$P^{\min}_g, P^{\max}_g$	Minimum and maximum power generation
$K^{\max}_{\text{line},\bar{l}}$	Maximum new line investment capacity
x_l	Base reactance of transmission line
$p_c(c)$	Probability of each case c happening
$P^{s,c}_g, P^{s,c}_{\text{loss},d}$	Power generation and amount of load being shed for scenario s of case c
$P^{s,c}_g, f^{s,c}_{\text{line}}, P^{s,c}_{\text{loss}}$	Vectors of power generation, transmission line flow, and amount of load being shed for scenario s of case c
$P^{\min}_{\text{loss},d}, P^{\max}_{\text{loss},d}$	Minimum and maximum amount of load being shed
$P(s'|S_t, a_t)$	State transition probability (i.e., probability of transitioning to state s' given state S_t and action a_t)
$R_a(s', s)$	Reward function giving the value of transitioning from state s to state s' after taking action a
$C_{\text{op}}(\{\cdot\})$	Annual operational cost at the state defined by $\{\cdot\}$
$C_{\text{op+invt}}(\{\cdot\})$	Annual operational cost and annualized investment cost at the state defined by $\{\cdot\}$
$R_{\text{term}}(s)$	The terminal reward for being at state s
i	Interest rate
N	Number of decision cycles
$\text{NC}_s, \text{ML}_s, \text{GC}_s$	Network configuration, maximum load level, and generation capacity defining state s
γ	Discount factor
$\pi(s)$	Policy function defining the action the decision maker will choose in state s at each time period t

| $V_\pi(s)$ | Expected discounted sum of cumulative reward over the time horizon if policy $\pi(s)$ is used |
| $v_t(s)$ | Value of being at state s at period t |

APPENDIX 8.B MATHEMATICAL MODELS USED FOR CASE STUDIES

Models for Optimal Transmission Investment Decision Making under Uncertainty

We developed two models to evaluate the optimal investment decision making in conventional transmission line and flexible reactance device. The first model accounts for the value of operational flexibility, while the second model extends the first model to account for the value of investment flexibility. Here, operational flexibility is the ability of the system to effectively react to short-run uncertainties and system conditions, whereas investment flexibility is the ability of the investment plan to react to long-run uncertainties and changes in system conditions. A three-step transmission investment decision-making framework is used for the models and summarized in Figure 8.B.1.

The first step of the framework is to characterize the uncertainties being considered. In order to select an appropriate method to characterize uncertainty, a sound understanding of the uncertainty being characterized and the information available is important. The investment decision is strongly influenced by which uncertainties are characterized and how they are characterized. In addition, it is also strongly affected by the institutional structure of the electricity industry. Factors such as who is making the investment decision and what factors are shared among stakeholders have a strong effect on the resulting investment decision.

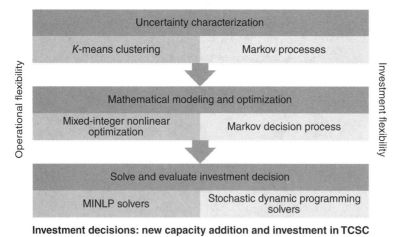

Figure 8.B.1   Flowchart of transmission investment decision-making framework.

The second step of the framework is to develop a mathematical model of the decision-making problem. The objective function and constraints of the mathematical models are dependent on how uncertainty is characterized and the operational framework that is being modeled. In addition, the decision-making problem can be modeled as either a static or a dynamic decision-making process. In a static decision-making process, a single optimal decision plan is found. In a dynamic decision-making process, the optimal decision plan will be dependent on information that is received over time.

The final step of the framework is to solve the mathematical model developed in order to evaluate the investment decision. The optimal investment decision-making problem can be difficult to solve for realistic problem size. New algorithms and heuristics will need to be developed for large-scale decision-making problem. Since the goal of this chapter is not to design new algorithms, the models are implemented for small case studies that can be solved using readily available optimization software and packages.

Model 1: Valuing Short-Term Operational Flexibility for Transmission Investment Decision

In this first model, the goal is to evaluate how the optimal investment decision making is affected by short-run operational uncertainties under different operational frameworks. Short-run uncertainties considered here include uncertainty in renewable energy and load generation and also potential outages in transmission elements.

Step 1: Uncertainty Characterization Using K-Means Clustering

For the purpose of this problem, we are interested in the influence of short-run fluctuations in wind and loads and transmission element outages to investment decision. Uncertainty in transmission element outages is captured by the probability that the given element will fail. This probability can be obtained from historical outage data for similar line elements or engineering knowledge. More effort was taken to properly characterize wind and load uncertainty.

Transmission line flow is affected not only by the level of wind and load at each bus, but also by the relative level of wind and load among the buses. Therefore, we select an uncertainty characterization method that allows us to capture not only the uncertainty in the wind and load level at a given bus, but also the correlation among the wind and load level of different buses. The K-means clustering algorithm was selected as it allows us to create a set of representative wind and load scenarios that accurately capture the wind and load profile at each bus as well as the correlation among the wind and load in the system. K-means clustering was used in [18] to create wind and load scenarios for wind generation investment decision.

K-means clustering is an iterative algorithm that clusters data into similar groups. In this case, the dataset is historical wind and load data for the entire system. Each data point is the wind and load data at all nodes for a given point of time. The K-means clustering algorithm is used to group the dataset into a representative set of clusters. The centroid of each cluster will give us the wind and load level for one scenario, and the number of observations in each cluster will give us the probability that the scenario happens. A general outline of the *K*-means algorithm is shown below [18]:

1. Select the appropriate number of clusters (*N*). One method of selecting the number of cluster is by testing out different numbers of cluster and plotting out the graph of percentage of variance explained by cluster versus the number of cluster. The appropriate number of cluster is selected such that any additional increase in the number of cluster does not produce any substantial increase in model performance.

2. Randomly select *N* points from the dataset to be used as initial centroid for the *N* clusters.

3. The squared Euclidean distances between each original data point and the centroids are calculated. Each original data point is assigned to the cluster that it has the shortest Euclidean distance to.

4. Calculate the mean of each cluster to be the new centroid.

5. Repeat steps 3–5 until there are no changes in cluster composition between iterations and store the clusters and the total sum of distance of the resulting clusters.

6. In order to reduce the possibility of landing in a local minimum, repeat steps 2–5 for a user-selected number of times.

7. Select the results with the minimum total sum of distance as the final clusters.

Step 2: Mathematical Modeling and Optimization for Optimal Investment

In this step, a mathematical model of the optimal investment problem is developed. The objective of the problem is to minimize the overall expected operational and investment cost of the system, subjected to operational and investment constraints. The exact forms of the objective function and constraints are dependent on the operational rules being modeled. The generic form of the mathematical optimization problem can be written as

$$\min_{x,y} \sum_{s=1}^{\text{NumScenario}} Ct(s)C_{\text{opt}}(x, s) + C_{\text{invt}}(y) \tag{8.B.1a}$$

$$\text{s.t.} \quad g_{\text{op}}(x, y, s) = 0 \tag{8.B.1b}$$

$$h_{\text{op}}(x, y, s) \leq 0 \tag{8.B.1c}$$

$$g_{\text{invt}}(y, s) = 0 \tag{8.B.1d}$$

$$h_{\text{invt}}(y, s) \leq 0 \tag{8.B.1e}$$

$$x_{\min} \leq x \leq x_{\max} \tag{8.B.1f}$$

$$y_{\min} \leq y \leq y_{\max} \tag{8.B.1g}$$

where *x* represents the operational decision variables, *y* represents the investment decision variables, C_{opt} represents the hourly operational cost, C_{invt} represents the annualized investment cost, and $Ct(s)$ is the number of hours scenario *s* happens in a year. Equations (8.B.1b) and (8.B.1c) represent the operational constraints, while equations (8.B.1f) and (8.B.1g) represent the investment constraints. The investment decision variables (y) include investment capacity for both new transmission lines and devices that provide flexible reactance.

Optimal investment decision using four different types of operational dispatch framework is modeled in this chapter:

- Economic dispatch with economic considerations only
- Preventive $N-1$ security-constrained economic dispatch
- Corrective $N-1$ security-constrained economic dispatch with inelastic load
- Corrective $N-1$ security-constrained economic dispatch with elastic load

The four different models are outlined next.

Economic Dispatch with Economic Considerations Only   The first case considered uses the standard economic dispatch where the only operational objective is to minimize generation cost. The objective function here is hence to minimize the expected generation cost and transmission investment cost:

$$\min_{\substack{P_g^s, x_{\text{Flex}}^s, \theta_n^s, f_{\text{line}}^s, \\ b_{\text{line}}, K_{\text{line}}, K_{\text{TCSC}}}} \sum_{s=1}^{N_s} Ct(s) \sum_{g=1}^{N_G} c_g \left(P_g^s\right) + \sum_{\bar{l}=1}^{N_{\bar{l}}} c_{\text{invt},\bar{l}}(K_{\text{line},\bar{l}}) + \sum_{t=1}^{N_T} c_{\text{invt},t}(K_{\text{Flex},t}) \quad (8.B.2a)$$

where $N_s, N_G, N_{\bar{l}}, N_T$ represent the number of scenarios, number of generators, number of new lines, and number of flexible devices, respectively; $c_g, c_{\text{invt},\bar{l}}, c_{\text{invt},t}$ represent the generation cost, new line investment cost, and new flexible reactance cost, respectively; P_g^s is the power generation; x_{Flex}^s is the operational setting of the flexible reactance device in terms of change in reactance; θ_n^s represents the nodal angles; f_{line}^s represents the line flows; $K_{\text{line}}, K_{\text{Flex}}$ represent new transmission line and flexible reactance capacity, respectively; and b_{line} is a binary variable indicating whether the new line is built.

The operational constraints are

$$GP_g^s + Sf_{\text{line}}^s - D^s = 0 \quad \text{for } \forall s \tag{8.B.2b}$$

$$P_g^{\min} \leq P_g^s \leq P_g^{\max} \quad \text{for } \forall s, g \tag{8.B.2c}$$

$$f_{\text{line},l}^s = \frac{\theta_{l,\text{to}}^s - \theta_{l,\text{from}}^s}{x_l - x_{\text{Flex},t}^s} \quad \text{for } \forall s, l \notin \bar{l} \tag{8.B.2d}$$

$$-K_{\text{line},l} \leq f_{\text{line},l}^s \leq K_{\text{line},l} \quad \text{for } \forall s, l \notin \bar{l} \tag{8.B.2e}$$

$$-M(1 - b_{\text{line},\bar{l}}) \leq f_{\text{line},\bar{l}}^s - \frac{\theta_{l,\text{to}}^s - \theta_{l,\text{from}}^s}{x_{\bar{l}} - x_{\text{Flex},t}^s} \leq M(1 - b_{\text{line},\bar{l}}) \quad \text{for } \forall s, \bar{l} \tag{8.B.2f}$$

$$-K_{\text{line},\bar{l}} \leq f_{\text{line},\bar{l}}^s \leq K_{\text{line},\bar{l}} \quad \text{for } \forall s, \bar{l} \tag{8.B.2g}$$

$$0 \leq x_{\text{Flex},t}^s \leq K_{\text{Flex},t} \quad \text{for } \forall s, t \tag{8.B.2h}$$

Equation (8.B.2b) is the nodal power balance, where G is a binary matrix indicating the nodal location of the generators, S is a matrix that is -1 for a transmission line exiting a node and $+1$ for a transmission line entering a node, and $P_g^s, f_{\text{line}}^s, D^s$ are vectors of power generation, transmission line flow, and nodal power demand, respectively. Equation (8.B.2c) represents the generation limit. Equations (8.B.2d) and (8.B.2e) are the line flow constraints for existing transmission lines, where $\theta_{l,\text{to}}^s, \theta_{l,\text{from}}^s$ are the nodal power angles of the existing and entering node of the transmission line, x_l represents the transmission line reactance, and the remaining variables are as defined earlier. Equations (8.B.2f) and (8.B.2g) are the line flow constraints for new transmission lines. Equation (8.B.2f) is a big M constraint, which enforces the line flow constraint only when the line is actually built. Finally, equation (8.B.2h) is the operational limits of the flexible reactance device.

The investment constraints are

$$0 \leq K_{\text{line},\bar{l}} \leq K_{\text{line},\bar{l}}^{\max} b_{\text{line},\bar{l}} \quad \text{for } \forall \bar{l} \tag{8.B.2i}$$

$$0 \leq K_{\text{Flex},t} \leq 0.5 x_l \quad \text{for } \forall l \tag{8.B.2j}$$

Equations (8.B.2i) and (8.B.2j) are the investment limits of new transmission lines and flexible reactance device, respectively. In this chapter, we assume that the maximum capacity of flexible reactance device that can be installed in a line is 50% of the base reactance of the line.

Preventive N − 1 Security-Constrained Economic Dispatch In the first case, we assume that power is dispatched by minimizing the generation cost only. In actual system operation, power is often dispatched more conservatively due to security constraints to maintain the reliability of the grid. In this second case, we assume that the power system is operated under $N - 1$ security. This means that power is dispatched such that the system will remain stable even if any single system element fails. Under preventive $N - 1$ security, the system should remain stable without any corrections in power dispatch or control settings in the system. For the purpose of this chapter, we only consider failures in transmission elements (i.e., transmission line or flexible reactance device).

The preventive $N - 1$ security-constrained economic dispatch is similar to the basic economic dispatch problem presented in the last section. The only difference is that there is an increase in the number of operational constraints. The base case set of operational constraints shown earlier is replicated with one transmission element removed at a time. Therefore, the final number of operational constraints is equal to (1 + the total number of transmission elements) times the number of operational constraints in the base case set of operational constraints shown earlier. For each scenario, the power dispatch and flexible reactance settings have to remain the same for all $N - 1$ contingencies; however, the line flows and nodal power angles will be different and governed by Kirchhoff's laws.

Corrective N − 1 Security-Constrained Economic Dispatch with Inelastic Load Preventive $N - 1$ operation results in conservative operation of the system that is often highly inefficient. In recent years, "smart grid" proponents have proposed the use of a corrective $N-1$ operational strategy to minimize the inefficiency in the system [11]. In corrective $N-1$ operation, corrective dispatch and control changes can be made to the system when outages happen as long as it can be done within a certain time frame. The main argument for corrective $N-1$ operation is that the system can often remain stable for a short time after a certain outage happens, and that corrective actions can be taken within the short time frame to stabilize the system. In this case, we assume that the potential corrective action that can be taken changes the power dispatch and flexible reactance control setting.

As with the previous case, the number of operational constraints is equal to (1 + the total number of transmission elements) times the number of operational constraints in the base case operational constraints. However, in this case, the power dispatch and control settings of the flexible reactance are allowed to change for each $N-1$. In addition, adjustment needs to be made to the objective function to account for the probability of each of the contingencies happening. The objective function for the corrective $N-1$ security-constrained economic dispatch is given by

$$\min_{\substack{P_g^{s,c}, x_{Flex}^{s,c}, \theta_n^{s,c}, f_{line}^{s,c}, \\ b_{line}, K_{line}, K_{Flex}}} \sum_{c=1}^{N_c} p_c(c) \sum_{s=1}^{N_s} ct(s) \sum_{g=1}^{N_G} c_g\left(P_g^s\right) + \sum_{\bar{l}=1}^{N_{\bar{l}}} c_{invt,\bar{l}}(K_{line,\bar{l}}) + \sum_{t=1}^{N_T} c_{invt,t}(K_{Flex,t})$$

(8.B.3)

where $p_c(c)$ is the probability of each case c happening. The cases include the base case with no contingencies and the cases representing all $N-1$ contingencies. The other variables are as defined earlier.

Corrective N − 1 Security-Constrained Economic Dispatch with Elastic Load   In the previous case, we assume that the load is inelastic and non-dispatchable. In this case, we assume that part of the load is elastic and dispatchable. A "loss-of-load" price is assigned to each load. This loss-of-load price is the amount that needs to be compensated to the consumer if their load is shed at any point of time. The new operational objective is hence to minimize both expected generation and load cost:

$$\min_{\substack{P_g^{s,c}, x_{Flex}^{s,c}, \theta_n^{s,c}, f_{line}^{s,c}, \\ b_{line}, K_{line}, K_{Flex}}} \sum_{c=1}^{N_c} p_c(c) \sum_{s=1}^{N_s} ct(s) \left[\sum_{g=1}^{N_G} c_g\left(P_g^{s,c}\right) + \sum_{d=1}^{N_D} c_D\left(P_{loss,d}^{s,c}\right) \right] + \sum_{\bar{l}=1}^{N_{\bar{l}}} c_{invt,\bar{l}}(K_{line,\bar{l}})$$

$$+ \sum_{t=1}^{N_T} c_{invt,t}(K_{Flex,t})$$

(8.B.4a)

where N_D is the number of dispatchable load, $P_{loss,d}^{s,c}$ is the amount of load being shed, c_D represents the price of loss load, and the remaining variables are as defined earlier.

The power balance constraints shown earlier equation (8.B.2b) needs to be altered to account for the potential of dispatching load:

$$GP_g^{s,c} + Sf_{line}^{s,c} D^s + P_{loss}^{s,c} = 0$$

(8.B.4b)

In addition, an additional constraint representing the capacity limit for the dispatchable load needs to be added:

$$P_{\text{loss},d}^{\min} \leq P_{\text{loss},d}^{s,c} \leq P_{\text{loss},d}^{\max} \quad \text{for } \forall\, s, d \tag{8.B.4c}$$

Step 3: Solve Optimization Problem to Evaluate Investment Decision

The optimization problems presented above are mixed-integer nonlinear programming problems (MINLP). MINLP are challenging problems to solve. Even though various optimization algorithms and heuristics have been developed in recent years to enable us to better solve such problems, large-scale MINLP are still difficult to solve. The goal of this chapter is not to design new algorithm and heuristics to solve the challenging optimal transmission investment decision problem. Instead, the goal is to use simple case studies to generate insights into institutional and policy design for flexible transmission technologies. The case studies used in this chapter can be solved by generic MINLP solvers such as BARON and SCIP that can be used with traditional optimization platforms. In this case, OptiToolbox for MATLAB is used to implement SCIP to solve the optimization problem [19].

Model 2: Valuing Long-Term Investment Flexibility for Transmission Investment Decision

In this second model, the goal is to evaluate how the optimal investment decision making is affected by different levels of long-run uncertainties and with different level of load responsiveness. The long-run uncertainties that are considered here include long-run uncertainties in load growth and generation investment patterns.

Step 1: Uncertainty Characterization Using Markov Processes

In this chapter, the uncertain load growth and generation investment patterns are modeled using a discrete-time Markov process (also known as Markov chain). A Markov process is a memoryless process, which means that the next state of the process is only dependent on the current state and not on any previous state. Markov chains are described by discrete states and state transition matrix. A graphical illustration of a hypothetical Markov chain for long-term load growth is shown in Figure 8.B.2.

In Figure 8.B.2, the nodes (L1, L2, L3, L4, L5) represent the five different states (i.e., load levels in this example). The numbers on the edges represent the state transition probabilities. The numbers on the self-loops tell us the probability that the state will stay the same in the next time period, while the numbers on the edges connecting two nodes tell us the probability of transitioning from one state to another. For instance, if the current load level is L1, there is a 50% chance that the next load level will be the same, a 25% chance that the next load level will be L2, and a 25% chance that the next load level will be L3.

The challenge of obtaining a Markov chain representation of uncertainty in generation and load growth lies in determining the appropriate state transition matrix. Historical system data and expert elicitation can be used to determine the appropriate state and state transition matrix for a Markov model.

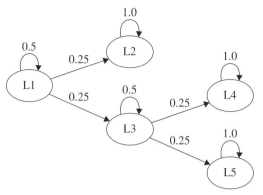

Figure 8.B.2   Graphical illustration of a hypothetical Markov chain for long-term load growth.

Step 2: Mathematical Modeling and Optimization for Optimal Investment

A stochastic dynamic programming is used to model this long-term optimal investment problem. In particular, the optimal investment decision problem is modeled as a discrete-time MDP. MDP provides an elegant framework to model decision making under uncertainty under a set of modeling assumptions [20]. These assumptions are as follows:

- There are a finite number of states.
- There are a finite number of actions/decisions.
- The system state transition process has the Markov property. Future states are only influenced by the current state and not previous states.
- The cost/reward associated with each state–action pair can be computed.

The five key components of an MDP problem formulation are (1) decision epoch, (2) state space, (3) action space, (4) transition probabilities, and (5) reward function [20]. The MDP problem formulation for the long-term optimal investment problem will be presented next.

Decision Epoch (t)   The decision epoch in our problem formulation represents the time interval in which investment decision is being made. This is dependent on the institutional framework governing investment decisions. An annual decision-making cycle would be a reasonable assumption to make here. However, to simplify computation, a 4-year decision cycle is used in this chapter.

State Space (S) and Action Space (A)   The states of the model are described by three types of state variables: {Network Configuration (NC), Maximum Load Level (ML), Generation Capacity (GC)}. The action space is the set of transmission investment options.

State Transition Probabilities $P(s'|S_t, a_t)$ The load level and generation capacity are modeled as Markov processes as described earlier. They are exogenous state variables and the state transition matrix is assumed to not be affected by the actions taken (i.e., $P(s'|S_t, a_t) = P(s'|S_t)$). The network configuration represents the state of the transmission network, including both new lines and flexible reactance device. This is an endogenous state variable where the state transition matrix is dependent on the decision being taken (i.e., $P(s'|S_t, a_t) \neq P(s'|S_t)$).

Reward Function $(R_a(s', s))$ The reward function gives the value of transitioning from state s to state s' after taking action a. The reward function represents the savings in operational cost in state s' given the new network configuration. It can be calculated using the following formula:

$$R_a(s', s) = C_{op}\left(\left\{NC_0, ML_{S'}, GC_{S'}\right\}\right) - C_{op+invt}\left(\left\{NC_{s'}, ML_{s'}, GC_{s'}\right\}\right) \quad (8.B.5a)$$

where $C_{op}(\{\cdot\})$ is the annual operational cost at the state defined by the state variables and $C_{op+invt}(\{\cdot\})$ is the annual operational cost and annualized investment cost at the state defined by the state variables. The first term of the formula represents the operational cost at the new load level and generation capacity assuming base case configuration. The second term in the formula represents the operational cost and total annualized investment cost at the new load level and generation capacity at the new network configuration. The operational cost can be obtained by running the optimal power flow for the set of scenarios obtained using the K-means clustering technique discussed earlier adjusted based on the appropriate network configuration, load level, and generation capacity.

Terminal Reward Since the problem is modeled as a finite horizon MDP, a terminal reward is defined for the problem. The terminal reward is defined as the expected operational savings due to the investment minus the cost of investment over the remaining lifetime of the investment. It is assumed that the state remains at the final state for the rest of the lifetime of the investment. The terminal reward is calculated based on the following formula:

$$R_{term}(s) = \frac{1 - (1 + i)^{-n}}{i} \left[C_{op}\left(\left\{NC_0, ML_S, GC_S\right\}\right) - C_{op+invt}\left(\left\{NC_s, ML_s, GC_s\right\}\right)\right]$$
$$(8.B.5b)$$

The first term in the above equation represents the operational cost at the final load and generation level over the remaining time if the network configuration remained at the base case (i.e., no investment is made). The second term represents the operational cost and annualized investment cost at the final load and generation level with the new network configuration.

Solution Objective The goal of the MDP problem is to determine the best "policy." The policy function $(\pi(s))$ specifies the action that the decision maker will choose in state s at each time period t. At every time period the decision maker can choose to do nothing and wait for more information, invest in TCSC, or invest in new line. The goal is to choose a policy that maximizes the expected discounted sum of cumulative reward over the time horizon:

$$V_\pi(s) = E_{\pi,s} \left\{ \sum_{t=1}^{T} \gamma^t R_\pi(s', s) \right\} \tag{8.B.5c}$$

where γ is the discount factor and R_π is the reward due to following the policy.

Step 3: Solve Optimization Problem to Evaluate Investment Decision

The MDP can be solved using various dynamic programming algorithms such as policy iteration, value iteration, and various approximate dynamic programming techniques. For a larger problem with a large state space, various heuristics and approximation techniques will be needed to fully solve the dynamic programming problem. However, for the purpose of this chapter, the basic backward induction technique is used to solve this finite-stage MDP. The backward induction algorithm is a recursive algorithm that does the following [21]:

1. For the final period T, let the value function at the final period be equal to the terminal reward:

$$v_T(s) = R_{\text{term}}(s) \tag{8.B.6a}$$

2. Set counter $n = 1$.
3. For $t = T - n$ and for all states s, calculate the following:

$$v_t(s) = \max_{a \in A} \left\{ \sum_{s' \in S} P_a(s', s)(R_a(s', s) + \gamma v_{t+1}(s')) \right\} \tag{8.B.6b}$$

$$\pi_t(s) = \arg\max_{a \in A} \left\{ \sum_{s' \in S} P_a(s', s)(R_a(s', s) + \gamma v_{t+1}(s')) \right\} \tag{8.B.6c}$$

4. Increase counter $n = n + 1$.
5. If $n = T$, end algorithm. Otherwise, repeat steps 2 to 5.

The MDP Toolbox in MATLAB is used to solve the case studies presented in this chapter [21].

APPENDIX 8.C INVESTMENT COST

Investment cost has a huge influence on the overall investment decision. The investment costs for new transmission capacity expansion and flexible reactance are calculated based on the following assumptions.

The transmission line cost was calculated using the following cost assumptions:

- Investment cost: $1000 per megawatt mile [3]
- Length of transmission line is 200 miles

TABLE 8.C.1 Annualized investment costs

	Annualized Investment Cost
New line investment	$23k/MW
Flexible reactance device in line 1	$270k/pu flexible reactance

- 10% interest rate
- 20-year lifetime of investment

The flexible reactance device investment cost was calculated using the following cost assumptions for TCSC:

- Investment cost: $135/kVAR [5]
- The MVar operating range (s) of TCSC can be calculated using the following formula [5]:

$$s = X_c \frac{K_{line}^2}{S_{base}} \qquad (8.C.1)$$

where X_c is the maximum series capacitor reactance in pu, K_{line} is the thermal line limit of the transmission line, and S_{base} is the MVA base power (100 MVA in this case).

- 10% interest rate
- 20-year lifetime of investment

The investment cost assumption for the transmission line used is the lower end of the cost given in the referenced source. The resulting investment costs used in this case study are listed in Table 8.C.1.

REFERENCES

[1] S. Cvijic, "A Modeling Framework for Capturing Parallel Flows and Effects of Bilateral Contracts in an Evolving Industry," Master Thesis, Carnegie Mellon University, Pittsburgh, PA, 2010.

[2] P. Joskow and J. Tirole, "Merchant Transmission Investment," National Bureau of Economic Research Working Papers 9534, March 2003.

[3] P. Donohoo-Vallett, "Design of Wide-Area Electric Transmission Networks under Uncertainty: Methods for Dimensionality Reduction," PhD Dissertation, Massachusetts Institute of Technology, Cambridge, 2014.

[4] American Electric Power, "Transmission Facts.". Available at www.slideshare.net/ashishkc007/transmission-facts [Accessed January 31, 2014].

[5] E. de Oliveira, J. W. Marangon Lima, and J. Pereira, "Flexible AC transmission system devices: allocation and transmission pricing," *International Journal of Electric Power and Energy Systems*, vol. 21, no. 2, pp. 111–118, 1999.

[6] R. Dunlop, R. Gutman, and P. Marchenko, "Analytical development of loadability characteristics for EHV and UHV transmission lines," *IEEE Transactions on Power Apparatus and Systems*, vol. 98, no. 2, pp. 606–616, 1979.

[7] National Research Council, *Electricity from Renewable Resources: Status, Prospects, and Impediments*, The National Academies Press, Washington, DC, 2010.

[8] Energy Information Administration, "Electric Power Monthly: Table 6.7.B. Capacity Factors for Utility Scale Generators Not Primarily Using Fossil Fuels, January 2008–November 2014," January 27, 2015. Available at http://www.eia.gov/electricity/monthly/epm_table_grapher.cfm?t= epmt_6_07_b [Accessed February 23, 2015].

[9] Federal Energy Regulatory Commission, "Uplift in RTO and ISO Markets," Federal Energy Regulatory Commission, Washington, DC, 2014.

[10] G. B. Alderete, "FTRs and revenue adequacy," in *Financial Transmission Rights*, Springer-Verlag, London, 2013, pp. 253–270.

[11] A. Monticelli, M. Pereira, and S. Granville, "Security-constrained optimal power flow with post-contingency corrective rescheduling," *IEEE Transactions of Power System*, vol. 2, no. 1, pp. 175–180, 1987.

[12] PJM, "PJM-Operational Analysis." Available at http://www.pjm.com/markets-and-operations/ops-analysis.aspx [Accessed February 1, 2015].

[13] L. Ghods and M. Kalantar, "Different methods of long-term electric load demand forecasting; a comprehensive review," *Iranian Journal of Electrical & Electronic Engineering*, vol. 7, no. 4, pp. 249–259, 2011.

[14] V. Sarkar and S. A. Khaparde, "A comprehensive assessment of the evolution of financial transmission rights," *IEEE Transactions on Power System*, vol. 23, no. 4, pp. 1783–1795, 2008.

[15] J. Arce and S. Wilson, "Managing Congestion Risk in Electricity Market," in Carnegie Mellon Conference on Electricity Transmission in Deregulated Markets: Challenges, Opportunities, and Necessary R&D Agenda, Pittsburgh, PA, 2004.

[16] J. Rosellon, "Mechanisms for the optimal expansion of electricity transmission networks," in *Financial Transmission Rights*,Springer-Verlag, London, 2013, pp. 191–210.

[17] B. Henze, C. Noussair, and W. Bert, "Long term financial transportation rights: an experiment," in *Financial Transmission Rights*, Springer-Verlag, London, 2013, pp. 211–226.

[18] I. Baringo and A. Conejo, "Correlated wind-power production and electric load scenarios for investment decision," *Applied Energy*, vol. 101, pp. 475–482, 2013.

[19] J. Currie and D. I. Wilson, "OPTI: Lowering the Barrier between Open Source Optimizers and the Industrial MATLAB User," Foundations in Computer-Aided Process Operations, Savannah, GA, 2012.

[20] O. Alagoz, H. Hsu, J. Schaefer, and M. Roberts, "Markov decision process: a tool for sequential decision making under uncertainty," *Medical Decision Making*, vol. 30, no. 4, pp. 474–483, 2010.

[21] I. Chades, M. Cros, F. Gracia, and R. Sabbadin, *Markov Decision Process Toolbox v2.0 for Matlab*, INRA, Toulouse, France, 2005.

INDEX

ACE data, 96–97. *See also* area control error (ACE)

ACE signals, 93–94, 97. *See also* CAISO ACE signal

action space (A), 245

active demand response, 62–63

adaptive robust optimization. *See also* robust optimization (RO)
  reserve determination via, 179–180
  stochastic programming vs., 181–184

aggregate model, 74–75

aggregation algorithms, 77–78

algorithms. *See also* solution algorithms
  aggregation, 77–78
  Benders decomposition, 136–137
  broadcast, 77
  ColorPower, 74–80, 82
  convex clearing, 34

alternative technologies, 220

ancillary service markets, 19–20

ancillary services (AS), 63–64, 72–73, 86–87, 92–93, 106, 108

angular stability, 8

architecture, ColorPower, 75–80

area control error (ACE), 5, 93–99. *See also* ACE entries
  calculating, 94

autocorrelation factor (ACF), 132

automatic generation control (AGC) system, 92–93

autoregressive moving average (ARMA) models, 132

awards, 86–87

balancing authority (BA), 93, 109

balancing authority area control error limit (BAAL) standard, 94
  regulation based on, 97–98

base case constraints, 135

basic dispatch model, for market clearing, 198–201

battery data, 156–162

Benders cut method, 125, 144–145, 148, 163, 172, 176

Benders decomposition (BD) algorithm, 136–137

bilinear heuristic, for robust reserve determination, 180–181

bilinear optimization problem, 147–148

binding transmission lines, 213
  at peak load, 214–216
  at valley load, 213

blackouts, 198

broadcast algorithms, 77

budget constraints, 142–143

business processes, 198–199

CAISO ACE signal, 94–96. *See also* California Independent System Operator (CAISO)

California "duck chart," 70–71

California electricity market, 45

California Independent System Operator (CAISO), 91–92
  day-ahead regulation forecast at, 93–99
  ramping and uncertainties evaluation at, 99–103

Power Grid Operation in a Market Environment: Economic Efficiency and Risk Mitigation, First Edition.
Edited by Hong Chen.
© 2017 by The Institute of Electrical and Electronics Engineers, Inc. Published 2017 by John Wiley & Sons, Inc.

candidate transmission lines, selecting, 206–210
capacity markets, 12–14, 59
capacity uncertainty, balancing, 101
capital formation, 48
case studies
    investment decision, 226–230
    long-term uncertainty, 230–234
    mathematical models for, 238–247
CCO-based SCUC, 151–155, 160–162. *See also* security-constrained unit commitment (SCUC)
central limit theorem, 151
chance-constrained optimization (CCO), 151–153. *See also* CCO-based SCUC
Clean Power Plan 2015, 133
clearing intervals, 60–61
clearing price, 43
clustering techniques, 227–229
collusion, 46
ColorPower
    algorithm/approach/method/system, 74–80
    control problem with, 77–79
    effectiveness of, 79–80
    uses for, 82
combined cycle unit modeling, 33
combustion turbine (CT) commitment, 22–24. *See also* CT optimizer (CTO)
common cause outages, 128–129
competitive constraints, 59–60
competitiveness, measuring, 46–47
competitive pressures, 58
competitive prices, 43–45
computational time, 190–191. *See also* CPU time
conduct and impact analysis, 47
conduct tests, 58–59
congestion rents (CRs), 206–207, 209–211, 213–216
constraint analysis, 23–24
consumer pricing models, 75
contingency criteria, 32–33
contingency evaluation subproblems, 123
contingency reserves, 6
control algorithms, ColorPower, 78–80
control architecture, 76–77
    ColorPower, 75–80

controllable capacitor, 221
control performance standards (CPSs), 94. *See also* CPS compliance
convex clearing algorithms, 34
correlation matrices, 228–229
CPS compliance, 96–97. *See also* control performance standards (CPSs)
CPU time, 124–125. *See also* computational time
CT optimizer (CTO), 25. *See also* combustion turbine (CT) commitment
cumulative distribution function (CDF), 101
current integration methods, problems with, 80–81
curtailment service providers (CSPs), 72
customer demand, 82–83

day-ahead (DA) markets (DAMs), 15–18, 49, 72, 80–81, 198
day-ahead regulation forecast, at CAISO, 93–99
day-ahead scheduling, 151, 153, 163, 169–193
DC model, in market clearing, 31–32
DC network constraints, 121, 125
DC power flow model, 204
decision epoch, 245
decision-making problem model, 239
demand, market power and, 83–84
demand bids, 50–51
demand flexibility information, aggregating, 74–75
demand forecast (load forecast), 5
demand increase, 231–232
demand response (DR), xv, 62–63, 69–70, 82. *See also* DR entries
    benefits from, 72
    generalization beyond, 84–87
    proxies to, 63–64
demand response management
    distributed computing-based, 74–75
    mass market, 69–88
demand-side bids, 60
demand willingness-to-pay, 44. *See also* willingness-to-pay
deterministic methods, xviii
    in market clearing, 32–33
deterministic reserve allocation models, 170

deterministic SCUC study, 157, 162–164.
See also security-constrained unit
commitment (SCUC)
dispatch, 26–28
dispatch actions, 21–22
dispatch errors, 94–96
dispatch time delay, 34–35
distributed computing-based demand
response management approach, 74–75
distributed energy sources (DERs), 69–70,
84
distributed resources, 85–86
distribution companies (DISCOs), 118
distribution facilities, 4
distribution system operators (DSOs),
70, 84
DR capacity, integrating with energy
markets, 80. *See also* demand response
(DR)
DR programs, 72–73, 81
current status of, 88
DR resources, 71, 73, 85
DR technologies, 71–72
"duck charts," 70–71
dynamic pricing, 80
dynamic programming (DP) method, 124
dynamic reserve requirements/zones, 20

economic dispatch, 226, 241–244
economic efficiency, xvii–xviii
balancing, 3–39
improving, 22–31
increasing, 33
economic efficiency
evaluation/improvement, 20–35
at PJM, 22–31
efficiency, achieving, 45. *See also* economic
efficiency entries; inefficiency;
inefficient entries; market efficiency
efficiency improvement, xv, 43–65
efficient dispatch, 43, 46
efficient operations, xvi
elastic load, 243–244
electrical devices, controlling, 77
electric energy systems, generation sources
for, 169
electricity demand, 4–5
electricity industry, 69
disconnects in, 225

electricity industry landscape, changes in,
219–220
electricity market model, 37–39
electricity market restructuring, 118
electricity markets, xvii, 10–11, 236
European wholesale, 16
global, 12–13
Latin America, 16
price formation in, 50–52
electricity restructuring, 91
electricity sector, actions needed in, 118
Electric Reliability Council of Texas
(ERCOT), 4, 7, 46, 59–61, 63
regulation service requirements at,
103–111
energy management systems (EMSs), 9–10,
93
energy market trading intelligence devices,
80
energy needs, flexibility in, 74
energy storage resources (ESRs), 69,
86–87
environmental issues, 118–119
European wholesale markets, 16
exhaustion rate, 107

facility thermal limitation, 7
feasibility analysis, 16–17
feasibility cuts, 150
Federal Energy Regulatory Commission
(FERC) compliance, 19–20, 72
financial rights, designing for flexibility,
234–235
financial transmission right (FTR) markets,
14–15
financial transmission rights (FTRs), 61,
216, 225
inefficiency of, 234–235
long-term, 235
5-bus test case, 210–213
flexibility
designing financial rights for, 234–235
incentives for, 220–228
in transmission planning, 219–248
flexibility services, 92
flexible reactance, 221–222
flexible reactance device, 223–224
flexible technologies, 219–221, 235–236
"flying brick" method, 102–103

forecasted load demand, 141
forecast errors, 94–96
forecast improvement, 29–31
forecasting methods, 30–31
forecasting provisions, 130–131
forecast uncertainties, 101
formulation properties, 172–175
forward contracts, 56–57
fully dispatched market participant, 47

generation, load and, 5–6
generation amounts, 170
generation balance, 5
generation companies (GENCOs), 118
generation curtailment amount, 171
generation excess, 173
generation investment, 231–232
generation regulation, 233
generation sources, 169
generator capacity constraints, 55–56
generator data, 155–156
generator input–output curves, 59
grid, uncertainties in, 220. *See also* leverage
    grid controllability; smart grid
grid reliability, xv
grid-scale energy storage, 120

heuristic-based methods, 124
hot-start capability, 125
hourly feasibility checks, 137–139
hourly network evaluation, 137, 140, 144
"hybrid" model, 16

IEEE Power Energy Society (PES) Task
    Force, xviii
IEEE 30-bus system, 185
impact tests, 58–59
independent system operators (ISOs), 50,
    91
independent system operators/regional
    transmission operators (ISOs/RTOs), 3,
    11, 21, 37, 72–73, 117
independent system operators/transmission
    system operators (ISOs/TSOs), 70–72.
    See also ISO/TSO markets
individual CCO (ICCO), 151. *See also*
    chance-constrained optimization (CCO)
individual evaluation, 202

inefficiency, economic significance of,
    48–49
inefficient dispatch, 46
inefficient production, 56
information exchange, 233
information sharing, 230–234
infrastructure/operational tools, 235
infrastructure planning, 225
integrated operation, 3, 11–12
integration methods
    new, 81–83
    problems with, 80–81
interchange forecast, 31
interconnected power systems, 4, 5
Internet communications, 70
Internet of Things (IoT), 69, 71
intra-day commitment/markets, 18–19
inverse residual demand functions, 54
investment cost, 247–248
investment decision case study, 226–230
investment decision evaluation, 244
investment decisions, 236
    evaluating, 244, 247
    uncertainty impact on, 230–234
investment flexibility, 235, 238
investment planning, operational and market
    practices on, 225–230
ISO/TSO markets, 71–73, 84–85, 87. *See
    also* independent system
    operators/transmission system
    operators (ISOs/TSOs)
IT-SCED, 24–26. *See also* SCED entries

Kirchhoff's current law (KCL), 221–222
K-means clustering technique, 227–229
    uncertainty characterization using,
    239–240

Lagrangian relaxation (LR) method, 35–36,
    124
large-scale power systems, 204–205
large-scale variable generation addition,
    91–113
largest minimum violation, 145–148
large system test case, 213–216
Latin American electricity markets, 16
Latin hypercube sampling (LHS), 135
Lerner index, 47

leverage grid controllability, optimal transmission switching and, 216
linear network model, 32
linear programming (LP) solutions, 35–36, 124, 125
linear sensitivity factor (LSF) method, 121, 125
linear violation constraints, 125
line power flow control devices, 221
LMP-based congestion management, 35–36, 39. *See also* locational marginal prices (LMPs)
load, generation and, 5–6
load data, 142–143
load demands, 143
load distribution factors (LDFs), 81
load elasticity, 232
load-following, 84
load forecast (demand forecast), 5, 129–130
load forecast uncertainty, 101
load-serving entities (LSEs), 81, 233–234
load shedding, 162–163
load uncertainties, 143
locational marginal prices (LMPs), 12, 18, 39, 81, 201. *See also* LMP entries
long-term FTRs, 235. *See also* financial transmission rights (FTRs)
long-term investment flexibility, for transmission investment decision, 244–247
long-term uncertainty case study, 230–234
look-ahead commitment and dispatch (LACD) process, 199
"loss-of-load" price, 232
loss-of-load probability (LOLP), 64, 151, 160–163
low-discrepancy techniques, 135
lower bound (LB), for sample average approximation, 176
lower buy bids, 82

Maintain network security, 7–9
marginal cost pricing, 20
marginal pricing, uniform, 37
market clearing
  basic dispatch model for, 198–201
  DC model in, 31–32
  deterministic methods in, 32–33
  resource modeling in, 33–34

market-clearing intervals, 60–61
market clearing model, 37
market-clearing prices, 50–51, 200–201
market concentration, 47
market design, 12–16
market design stages, 35–36
market efficiency, improving, 43–65, 197–217
market mechanisms, 225–226
  developing, 235
market operation, xvii, 3
market participants, 37
  motivations of, 53
market power, 44–46, 54
  ability and incentive to exercise, 53–57
  equalizing supply and demand for, 83–84
  measuring, 46–47
market power analysis, 49
market power mitigation, 43–46, 48–49, 65
  approaches to, 57–64
market prices, 20
market signals, 12
Markov chains, 132–133, 153
Markov decision processes (MDPs), uncertainty characterization using, 244–245. *See also* MDP problem formulation
Markov models, with ColorPower, 76
mark-ups, 55–56
mass market demand response management, 69–88
master UC problem, 137, 144, 150. *See also* unit commitment entries
mathematical modeling/models
  for case studies, 238–247
  for optimal investment, 245–247
MDP problem formulation, 245–247. *See also* Markov decision processes (MDPs)
minimum run time (MinRunTime), 25
mixed-integer nonlinear programming problems (MINLPs), 244
mixed integer programming (MIP) model, 124–125, 199–200
multiple binding transmission constraints, 54
multi-settlement energy market, 35–36

$N - 1$ security-constrained economic dispatch, 242–244

natural gas/electricity infrastructure, interdependence of, 133–134
natural water inflow, 132
net-load, 109
network topology control, xviii
nodal market model, 12, 14
nodal power balance, 242
nodal price differences, 61
nomenclature listings, 36–37, 89, 112–113, 164–166, 192–193, 216–217, 236–238
non-competitive constraints, 59–60
nongeneration resource (NGR) model, 86
nonspinning reserve service, 106
North America Electric Reliability Corporation (NERC), 6–7

offer-based economic dispatch, 50
offer-based security-constrained economic dispatch, 50, 53
offer caps, 52–53, 58
operating reserves demand curve (ORDC), 64
operational flexibility, 235, 238
    for transmission investment decisions, 239–244
operational/infrastructure tools, 235
operational mechanisms, 225–226
operational uncertainties, xvii–xviii. *See also* uncertainties
operation risk mitigation, xvii–xviii
    balancing, 3–39
    power system, 4–10
optimal investment, 51
    modeling and optimization for, 240–244
optimality checks, 139–140
optimal solutions, 178
    stochastic, 186–187
optimal transmission investment decision-making models, 238–239
optimal transmission switching (OTS), 203–206. *See also* OTS entries
    candidate transmission lines and, 206–210
    implementation of, 209–210
    leverage grid controllability and, 216
optimization, for optimal investment, 245–247
optimization methods/techniques, 124–125, 137

OTSBranch (OTSBr), 205–206, 209–210. *See also* optimal transmission switching (OTS)
OTS test cases, 210–216
outage-caused congestion exposure (OCCE), 201–202
outage evaluation, improving market efficiency via, 197–217
outage simulation process, 127
outer approximation (OA) method, 147
out-of-market (OOM) actions, 12, 20, 32

parameter changing, limitations on, 60–61
partial autocorrelation factor (PACF), 132
passive demand response, 62–64
pay-as-bid prices, 51
PCCR (production cost/congestion rent) measure, 209–210, 213–216
PD analysis, 26. *See also* perfect dispatch (PD)
PDF (probability density function), 101
PD performance score, 25
PD savings, 21–23
peak load, binding transmission lines at, 214–216
Pennsylvania–New Jersey–Maryland (PJM), 3
    economic efficiency improvement at, 22–31
    perfect dispatch at, 21–22
perfect dispatch (PD), 3. *See also* PD entries
    at PJM, 21–22
performance-based capacity markets, 14
power balance, 4–7
Power Energy Society (PES), xviii
power markets, transforming, 83
power system balance constraint, 141
power system balancing processes, 93
power system operation risk mitigation, 4–10
power system operations, 100
    challenges to, 197–198
    reserves for, 169–170
    trends in, 10
power system overview, 4
power systems, interconnected, 4–5
power system security, 117–118
power system topology, 198
power system uncertainties, 125–134

power transfer distribution factor (PTDF), 204

price caps, 52–53, 58

price formation, in electricity markets, 50–52

price mark-up, 47

price signals, 35

pricing mechanisms, 34

production costs (PCs), associated with transmission lines, 207–211, 213

profit maximization, 55–56
  analysis of, 53–54

pump storage modeling, 33–34

queuing theory, 71

ramping, at CAISO, 99–103

ramping capability, 133

ramping constraints, 136

ramping requirements, 103, 105

ramping uncertainties, 102

random generation amount, 171

random outages, of system components, 126–129

real-time bidding, 60

real-time commitment, 34

real-time energy market (RTM), 72, 199

real-time (RT) markets, 15–18, 49, 57. *See also* RT entries

real-time prices, 57

real-world customers, 73

reference prices, 60

reformulation method, for robust reserve determination, 180–181

regional transmission operators/independent system operators (RTOs/ISOs), *see* independent system operators/regional transmission operators (ISOs/RTOs)

regional transmission organizations (RTOs), 170, 197, 200

regulation, 92–93
  based on BAAL standard, 97–98

regulation data, 96–97

regulation requirements curve, 98–99

regulation service (RGS) requirement, 106–108, 111–112
  at ERCOT, 103–111
  wind generation effect over, 108–111

regulation services, 107–108

regulation up/down, 106–107

relative solar energy, 185–186

reliability assessment commitment (RAC) processes, 15, 22, 29, 31

reliability unit commitment (RUC) process, 199

renewable energy, 133–134, 171, 173–174, 185, 187, 191

renewable energy resources, xvi, 91, 131–133

renewable energy sources (RES), 69, 84

renewable generation, 117, 170, 198

reserve clearing prices (RCPs), 19

reserve commitments, xvi

reserve costs, 170

reserve determination
  via adaptive robust optimization, 179–180
  solution algorithm for, 180–181
  via stochastic programming, 170–179

reserve determination formulation, 170–172

reserve determination model, 171
  risk-neutral, 177

reserve determination/valuation, in day-ahead scheduling, 169–193

reserve markets, 19–20

reserves, 6–7
  for power system operation, 169–170
  processing, 191–192

reserve total costs, 189

reserve valuation, 185–191
  as a function of stochastic unit integration level, 186–188
  as a function of stochastic unit variability, 188–189

resource identity (resource ID), 85

resource modeling, in market clearing, 33–34

resource uncertainty, in SCUC, 134–155

responsive reserve, 106

revenue inadequacy, 234

reward function, 246–247

risk mitigation, *see* operation risk mitigation

risk-neutral reserve determination model, 177

risk-neutral stochastic model, solution algorithm for, 175–176

risk sharing, 230–234

RO-based SCUC, 140–150, 154–155, 159–160, 162. *See also* robust optimization (RO); security-constrained unit commitment (SCUC)
robust models, 189
  comparison of, 187
  formulation of, 181–182
robust optimal solutions, 187–189
robust optimization (RO), 140. *See also* adaptive robust optimization; RO-based SCUC
robust reserve determination, solution algorithm for, 180–181
robust reserve determination formulation, 179–180
robust reserve determination model, 181
RT operation, 21–22, 25–26. *See also* real-time entries
RT-SCED, 24, 26. *See also* SCED entries
RT unit commitment, 21, 34. *See also* unit commitment

sample average approximation (SAA) method, 175–176, 183
scale economies
  changes in, 225
  of transmission technologies, 221–225
scarcity pricing mechanism, 46
SCED process, 109. *See also* security-constrained unit commitment/security-constrained economic dispatch (SCUC/SCED) processes
SCED solutions, 35
scenario analysis, 202
scenario constraints, 136–140
scenario generation methods, 134–135
scenario reduction techniques, 135
SCUC constraints, 120. *See also* security-constrained unit commitment (SCUC)
SCUC formulation, 123
SCUC models, 118–123
SCUC solution methods, 123–125
seasonal/diurnal trends, 108
security-constrained economic dispatch (SCED) process, *see* SCED entries; security-constrained unit commitment/security-constrained economic dispatch (SCUC/SCED) processes
security-constrained unit commitment (SCUC), 119–125
  CCO-based, 151–155, 160–162
  deterministic, 157, 163–164
  implementing, 119
  resource uncertainty in, 134–155
  RO-based, 140–150, 154–155, 159–160, 162
  SP-based, 134–140, 154–155, 157–159, 162
  with uncertainties, 117–166
security-constrained unit commitment/security-constrained economic dispatch (SCUC/SCED) processes, 16–18. *See also* IT-SCED; RT-SCED; SCED entries
security constraints, 164, 191
security criteria, 32–33
security evaluation, 145
"service level" plans, 75
short-term economic assessment, 202
short-term forecasts, 63
short-term load forecasting, 129
simulation test, 87–88
six-bus system, 155
smart grid, 69, 71, 74, 118–119, 198
solar/wind generation, predicting impact of, 111
solar/wind generation load uncertainties, evaluating, 111
solution algorithms
  for risk-neutral stochastic model, 175–176
  for robust reserve determination, 180–181
solution objective, of MDP problem, 246–247
SP-based SCUC, 134–140, 154–155, 157–159, 162. *See also* security-constrained unit commitment (SCUC)
stability limitations, 8–9
standard market design (SMD), 118
state duration sampling, 127–128
state estimator (SE) solutions, 15–16
state sampling, 126–128
state space (S), 245
state transition matrix, 231

state transition probabilities, 246
state transition sampling, 127–128
stochastic models, 188
  comparison of, 186
stochastic optimal solutions, 186–189
stochastic production units, 169–170
stochastic programming (SP), 134–140,
    157–159, 170. *See also* SP-based SCUC
  adaptive robust optimization *vs.*,
    181–184
  reserve determination via, 170–179
stochastic units
  integration level of, 186–188
  variability of, 188–189
storage resources, 85–86
structural tests, 59
sub-resources, 85–86
supply, market power and, 83–84
supply-following, 84
supply-side mitigation approaches, 57–64
supply-side techniques, 65
surplus revenue, 234
sustainable energy system development, 10
synchronized reserves, 6
system components, random outages of,
    126–129
system load forecast inaccuracy, 129–131
system operation(s), xvii, 3
  challenges to, 111
  improving market efficiency in, 197–217
  uncertainties in, 27–29, 31–33
system operation risk mitigation, 4–10
system performance, improving, 91–113
system stability limitation, 8–9
system topology, 34

"take-it-or-leave-it" monopolistic price, 84
terminal reward, 246
Texas two-step, 59–60
thermal capacity limit, 222
thermal constraints, 64
thermal generating unit redispatch
    constraints, 122
thermal limitations, 7
thermal ratings, 7
thermal surrogate, 32
30-bus system, 185
three pivotal supplier (TPS) test, 24
time inflexibility, 75

time series models, 132
time-series wind/load modeling, 108–109
time-to-start (TTS), 25
tool capability case studies, 103
topology control, xviii
topology flexibility, 34
total system production cost, 20
transmission capacity, 224
transmission companies (TRANSCOs), 118
transmission constraints, 48, 51–52
  multiple binding, 54
transmission facilities, 4
transmission flexibility, 220
transmission flow, 207–208
transmission investment, 225
  requirements for, 232–233
transmission investment decisions
  long-term investment flexibility for,
    244–247
  operational flexibility for, 239–244
transmission line investment, 222–224
transmission line overload probability
    (TLOP), 160–163
transmission lines, 206–210, 242. *See also*
    binding transmission lines; violated
    transmission lines
transmission losses, 32
transmission network constraints, 135–136,
    141–142
transmission networks, 91
transmission network security constraints,
    122
transmission network violations, 150
transmission outages, 198, 201–202
  economic evaluation of, 201–203
transmission planning, flexibility in,
    219–248
transmission security constraints, 64
transmission security violations, 150
transmission switching, improving market
    efficiency via, 197–217. *See also*
    optimal transmission switching (OTS)
transmission switching candidates, selection
    method for, 209–210
transmission system operators (TSOs), 91
transmission technologies, scale economies
    of, 221–225
two-bus base case system, 222
two-bus power system, 176–179

UC solutions, 150. *See also* unit
  commitment entries
uncertainties. *See also* operational
  uncertainties
  characterizing, 227, 238
  in emerging power systems, 125–134
  information/risk sharing and, 230–234
  security-constrained unit commitment
    with, 117–166
  in system operation, 27–29, 33
uncertainties evaluation, at CAISO, 99–103
uncertainty characterization
  using *K*-means clustering technique,
    239–240
  using Markov processes, 244–245
uniform marginal pricing, 37
unit commitment, 20–21, 118, 120. *See also*
  RT unit commitment
unit commitment (UC) problems/solutions,
  123–125, 143–144, 150–151
unit performance, 27
unit-specific offer caps, 60
upper bound (UB), for sample average
  approximation, 176

valley load, binding transmission lines at,
  213
value of loss load (VOLL), 53, 64, 162
VAR% values, 188–189

violated transmission lines, 206, 213
virtual bids, 15, 49
voltage collapse, 32
voltage limitation, 9
voltage rating, 223
voltage stability, 8–9
voluntary mitigation plan, 61

water inflow, 132
weather-related outages, 129
Weibull random variables, 178–179
wholesale energy market, integration with,
  80–83
willingness-to-pay, 44, 47, 51, 62–63, 232
wind capacity, 109
wind data, 107, 142–143
wind generation effects, 108–111
wind generations, 101, 103–106, 143
wind-impact coefficiencies, 109–110
wind power generation, 132–133, 141
wind power generation forecasting tools,
  100
wind/solar generation, predicting impact of,
  111
wind/solar generation load uncertainties,
  evaluating, 111
wind uncertainties, 143

year-to-date (YTD), 25, 30

IEEE Press Series
on Power Engineering

Series Editor: M. E. El-Hawary, Dalhousie University, Halifax, Nova Scotia, Canada

The mission of IEEE Press Series on Power Engineering is to publish leading-edge books that cover the broad spectrum of current and forward-looking technologies in this fast-moving area. The series attracts highly acclaimed authors from industry/academia to provide accessible coverage of current and emerging topics in power engineering and allied fields. Our target audience includes the power engineering professional who is interested in enhancing their knowledge and perspective in their areas of interest.

1. *Principles of Electric Machines with Power Electronic Applications, Second Edition*
M. E. El-Hawary

2. *Pulse Width Modulation for Power Converters: Principles and Practice*
D. Grahame Holmes and Thomas Lipo

3. *Analysis of Electric Machinery and Drive Systems, Second Edition*
Paul C. Krause, Oleg Wasynczuk, and Scott D. Sudhoff

4. *Risk Assessment for Power Systems: Models, Methods, and Applications*
Wenyuan Li

5. *Optimization Principles: Practical Applications to the Operations of Markets of the Electric Power Industry*
Narayan S. Rau

6. *Electric Economics: Regulation and Deregulation*
Geoffrey Rothwell and Tomas Gomez

7. *Electric Power Systems: Analysis and Control*
Fabio Saccomanno

8. *Electrical Insulation for Rotating Machines: Design, Evaluation, Aging, Testing, and Repair, Second Edition*
Greg Stone, Edward A. Boulter, Ian Culbert, and Hussein Dhirani

9. *Signal Processing of Power Quality Disturbances*
Math H. J. Bollen and Irene Y. H. Gu

10. *Instantaneous Power Theory and Applications to Power Conditioning*
Hirofumi Akagi, Edson H. Watanabe, and Mauricio Aredes

11. *Maintaining Mission Critical Systems in a 24/7 Environment*
Peter M. Curtis

12. *Elements of Tidal-Electric Engineering*
Robert H. Clark

13. *Handbook of Large Turbo-Generator Operation and Maintenance, Second Edition*
Geoff Klempner and Isidor Kerszenbaum

14. *Introduction to Electrical Power Systems*
Mohamed E. El-Hawary

15. *Modeling and Control of Fuel Cells: Distributed Generation Applications*
M. Hashem Nehrir and Caisheng Wang

16. *Power Distribution System Reliability: Practical Methods and Applications*
Ali A. Chowdhury and Don O. Koval

17. *Introduction to FACTS Controllers: Theory, Modeling, and Applications*
Kalyan K. Sen and Mey Ling Sen

18. *Economic Market Design and Planning for Electric Power Systems*
James Momoh and Lamine Mili

19. *Operation and Control of Electric Energy Processing Systems*
James Momoh and Lamine Mili

20. *Restructured Electric Power Systems: Analysis of Electricity Markets with Equilibrium Models*
Xiao-Ping Zhang

21. *An Introduction to Wavelet Modulated Inverters*
S.A. Saleh and M.A. Rahman

22. *Control of Electric Machine Drive Systems*
Seung-Ki Sul

23. *Probabilistic Transmission System Planning*
Wenyuan Li

24. *Electricity Power Generation: The Changing Dimensions*
Digambar M. Tigare

25. *Electric Distribution Systems*
Abdelhay A. Sallam and Om P. Malik

26. *Practical Lighting Design with LEDs*
Ron Lenk and Carol Lenk

27. *High Voltage and Electrical Insulation Engineering*
Ravindra Arora and Wolfgang Mosch

28. *Maintaining Mission Critical Systems in a 24/7 Environment, Second Edition*
Peter Curtis

29. *Power Conversion and Control of Wind Energy Systems*
Bin Wu, Yongqiang Lang, Navid Zargari, and Samir Kouro

30. *Integration of Distributed Generation in the Power System*
Math H. Bollen and Fainan Hassan

31. *Doubly Fed Induction Machine: Modeling and Control for Wind Energy Generation Applications*
Gonzalo Abad, Jesus Lopez, Miguel Rodrigues, Luis Marroyo, and Grzegorz Iwanski

32. *High Voltage Protection for Telecommunications*
Steven W. Blume

33. *Smart Grid: Fundamentals of Design and Analysis*
James Momoh

34. *Electromechanical Motion Devices, Second Edition*
Paul C. Krause, Oleg Wasynczuk, and Steven D. Pekarek

35. *Electrical Energy Conversion and Transport: An Interactive Computer-Based Approach, Second Edition*
George G. Karady and Keith E. Holbert

36. *ARC Flash Hazard and Analysis and Mitigation*
J. C. Das

37. *Handbook of Electrical Power System Dynamics: Modeling, Stability, and Control*
Mircea Eremia and Mohammad Shahidehpour

38. *Analysis of Electric Machinery and Drive Systems, Third Edition*
Paul Krause, Oleg Wasynczuk, S. D. Sudhoff, and Steven D. Pekarek

39. *Extruded Cables for High-Voltage Direct-Current Transmission: Advances in Research and Development*
Giovanni Mazzanti and Massimo Marzinotto

40. *Power Magnetic Devices: A Multi-Objective Design Approach*
S. D. Sudhoff

41. *Risk Assessment of Power Systems: Models, Methods, and Applications, Second Edition*
Wenyuan Li

42. *Practical Power System Operation*
Ebrahim Vaahedi

43. *The Selection Process of Biomass Materials for the Production of Bio-Fuels and Co-Firing*
Najib Altawell

44. *Electrical Insulation for Rotating Machines: Design, Evaluation, Aging, Testing, and Repair, Second Edition*
Greg C. Stone, Ian Culbert, Edward A. Boulter, and Hussein Dhirani

45. *Principles of Electrical Safety*
Peter E. Sutherland

46. *Advanced Power Electronics Converters: PWM Converters Processing AC Voltages*
Euzeli Cipriano dos Santos Jr. and Edison Roberto Cabral da Silva

47. *Optimization of Power System Operation, Second Edition*
Jizhong Zhu

48. *Power System Harmonics and Passive Filter Designs*
J. C. Das

49. *Digital Control of High-Frequency Switched-Mode Power Converters*
Luca Corradini, Dragan Maksimovic, Paolo Mattavelli, and Regan Zane

50. *Industrial Power Distribution, Second Edition*
Ralph E. Fehr, III

51. *HVDC Grids: For Offshore and Supergrid of the Future*
Dirk Van Hertem, Oriol Gomis-Bellmunt, and Jun Liang

52. *Advanced Solutions in Power Systems: HVDC, FACTS, and Artificial Intelligence*
Mircea Eremia, Chen-Ching Liu, and Abdel-Aty Edris

53. *Power Grid Operation in a Market Environment: Economic Efficiency and Risk Mitigation*
Hong Chen